Student's Book 2

Authors:
Terry Hudson and Debbie Roberts

Advisor team:
Pamela Hunte • Amaala Muhammad
• Motiellal Singh

Macmillan Education Limited
4 Crinan Street
London N1 9XW

Companies and representatives throughout the world

ISBN 978-1-786-32564-8

First published 2016

Designed by Lucy Allan
Layout and typesetting by Blue Dog Design Studio
Illustrated by Kate Daubney c/o Advocate; Russ Daff, Niall Harding, Robin Lawrie, Richard Hoit and Nathalie Ortega c/o Beehive; Beccy Blake and Tek-Art
Cover design by Macmillan Education Limited
Cover illustration by Vladimir Aleksic
Picture research by Catherine Dunn

The publishers would like to thank the following education professionals for their valuable contributions:
Pamela Hunte (Barbados), Amaala Muhammad (St Vincent) and Motielall Singh (St Lucia)

The authors and publishers would like to thank the following for permission to reproduce their photographs:

Alamy/age footstock pp69(frog), 86(tr), Alamy/Ange p119(cl), Alamy/Art Directors & TRIP p49, Alamy/Cultura Creative (RF) p10(br), Alamy/Michael Dwyer p69(cow), Alamy/FLPA pp14(c), 113(tl), Alamy/Greatstock p121(bl), Alamy/Mike Hesp p88, Alamy/Images & Stories p23, Alamy/Roland Knauer pp104, 105, 108(bl), 110(br), Alamy/Paul Rapson p107(trr), Alamy/Borges Samuel p13(tr), Alamy/Tom Uhlman p69(mongoose), Alamy/Zoonar GmbH p121(tr); Getty pp30(cl), 69(shark), Getty/iStock/Konovalikov Andrey p95, Getty/DigitalVision/Yuri_Arcurs p35(tl), Getty/Caiaimage/Martin Barraud p83, Getty/iStock/Harald Biebel p101(crl), Getty/iStock/Susan Chiang p121(cr), Getty/Blend Images/Stewart Cohen p84(cr), Getty/Photolibrary/Sylvain Cordier p16, Getty/ Dirk Ercken Photography p71(frog), Getty/Photographer's Choice, RF/Georgette Douwma p24(1), Getty/Photodisc/Noel Hendrickson p18, Getty/Moment/Jens Karlsson p71(bc), Getty/EyeEm/Nuridin Kaliyev p86(tr insert), Getty/Moment/Brian E. Kushner p61(tr),

Getty/National Geographic/Hannele Lahti p73, Getty/Corbis Documentary/Frans Lemmens pp10(bl), 14(b), Getty/EyeEm/Cezary Lopacinski p61(tl), Getty/Photonica/Silvia Otte p50(tl), Getty/E+/s-cphoto p42(tl), Getty/Science & Society Picture Library p113(bc), Getty/Corbis/George Shelley p84(cl), Getty/E+/stockcam pp107(bc), 110(cl), Getty/Stocktrek Images p49(insert), Getty Images/iStockphoto Thinkstock Images p35(tr), Getty/Minden Pictures/NiS/Martin Woike p25, Getty/Federico Veronesi p70, Getty/EyeEm/Justin Wilson p108(br), Getty/EyeEm/Zumira Lima pp24(3), 27(c); **PHOTODISC** pp14(tr), 101(crr), PhotoDisc/Getty Images pp27(bl), 30(c); **Rex Features**/robertharding/Michael Runkel p107(cl), Rex Features/Richard Saker p92(bc);
Science Photo Library/British Antartic Survey p42(bl), Science Photo Library/Visuals Unlimited Inc/David Fleetham p100;
Shutterstock/alfotokunst p27(cr), Shutterstock/bouybin p45(br), Shutterstock/Willyam Bradberry p95, Shutterstock/Aleksandr Bryliaev p31(c), Shutterstock/MOHAMED TAZI CHERTI p97, Shutterstock/Marcel Clemens p61(bl), Shutterstock/T.Dallas p119(bl), Shutterstock/divedog p27(cl), Shutterstock/Gallinago_media p27(bc), Shutterstock/Great Teacher Thawatchai 50(cr), Shutterstock/Lindsay Helms pp14(cl), Shutterstock/ID1974 p48, Shutterstock/ Italianvideophotoagency p107(trl), Shutterstock/johnfoto18 p106(cr), Superstock/Wolfgang Kaehler pp10(c), 14(a), Shutterstock/Patryk Kosmider p106(cl), Shutterstock/Andrey_Kuzmin p71(bee), Shutterstock/Kae Deezign p32, Shutterstock/Kotzur Yang Creative p30(bc), Shutterstock/Brian Lasenby p27(tr), Shutterstock/Lerche&Johnson p42(cr), Shutterstock/MaraZe p45(tr), Shutterstock/Rob Marmion p84(tl), Shutterstock/Alex Mit p105(br), Shutterstock/Monkey Business Images p91, Shutterstock/Ruth Peterkin p74, Shutterstock/PHB.cz (Richard Semik) p105 (tr), Shutterstock/ Polryaz p117, Shutterstock/ppart p106(c), Shutterstock/Ondrej Prosicky pp24(4), 69(bird), Shutterstock/Rawpixel.com p92(c), Shutterstock/Paul Reeves Photography p24(2), Shutterstock/rukxstockphoto pp107(bl), 110(cr), Shutterstock/Eugene Sergeev p30(bl), Shutterstock/tale p113(br), Shutterstock/teena137 p33, Shutterstock/Thirteen p31(cl).

Printed and bound by CPI Group (UK) Ltd Croydon CR0 4YY

2022

10 9 8 7 6 5

Contents

How to use this book 4
Scope and sequence 6

Unit 1 Structure and function 10
- Plants and animals develop in different ways 12
- Science projects 20
- Check your learning so far 21

Unit 2 Ecosytems 1 22
- Adaptation of organisms to their environment 24
- Science projects 26
- Check your learning so far 27

Unit 3 Matter and materials 28
- Materials and their uses 30
- Designing an object 34
- Solids and liquids 40
- Water and ice 42
- Liquids at home 44
- Science projects 46
- Check your learning so far 47

Unit 4 The Earth's weather 48
- Measuring the weather 50
- Science projects 58
- Check your learning so far 59

Unit 5 The solar system 60
- The Sun, Earth and Moon 62
- Science projects 66
- Check your learning so far 67

Unit 6 Ecosystems 2 68
- Feeding relationships and defence in animals 70
- The effects of environmental destruction 72
- Solid waste management 75
- Making different sounds and musical instruments 78
- Science projects 80
- Check your learning so far 81

Unit 7 Forces, motion and structures 82
- Effects of forces 84
- Simple mechanical devices 86
- Science projects 88
- Check your learning so far 89

Unit 8 Diversity and classification 90
- Human variation 92
- Living and non-living things 94
- Characteristics of living things 96
- How do plants differ? 98
- Science projects 102
- Check your learning so far 103

Unit 9 Energy 104
- Making use of devices 106
- Windmills and waterwheels 108
- Science projects 110
- Check your learning so far 111

Unit 10 The Earth's resources 112
- Dealing with pollution 114
- Science projects 122
- Check your learning so far 123

Glossary 124

How to use this book

Mission: Science follows a topic- and scientific enquiry-driven approach modelled on the Grades K–6 OECS curriculum for primary science and technology. Develop your pupils' scientific and technological knowledge and understanding by using the Student's Book, and the accompanying Workbook and Teacher's Resources.

The Scope and sequence chart, on pages 6–9, can be used to support your short- and long-term planning, which will enable you to plan your progress through the OECS curriculum and easily identify each unit's scientific enquiries.

Each unit in Student's Book 2 develops and integrates the pupils' skills of enquiry within the unit topics that deliver the unit's learning outcomes. There are ten units which follow the curriculum areas for Grade 2 in the order in which they appear.

Unit opener

The unit opens with two pages of key words, ideas, facts and talking points to stimulate and engage pupils in the unit's main ideas before they begin exploring the topics contained within the unit.

Pupils are encouraged to discuss and share as a class, in groups or in pairs their knowledge and understanding of an idea or topic.

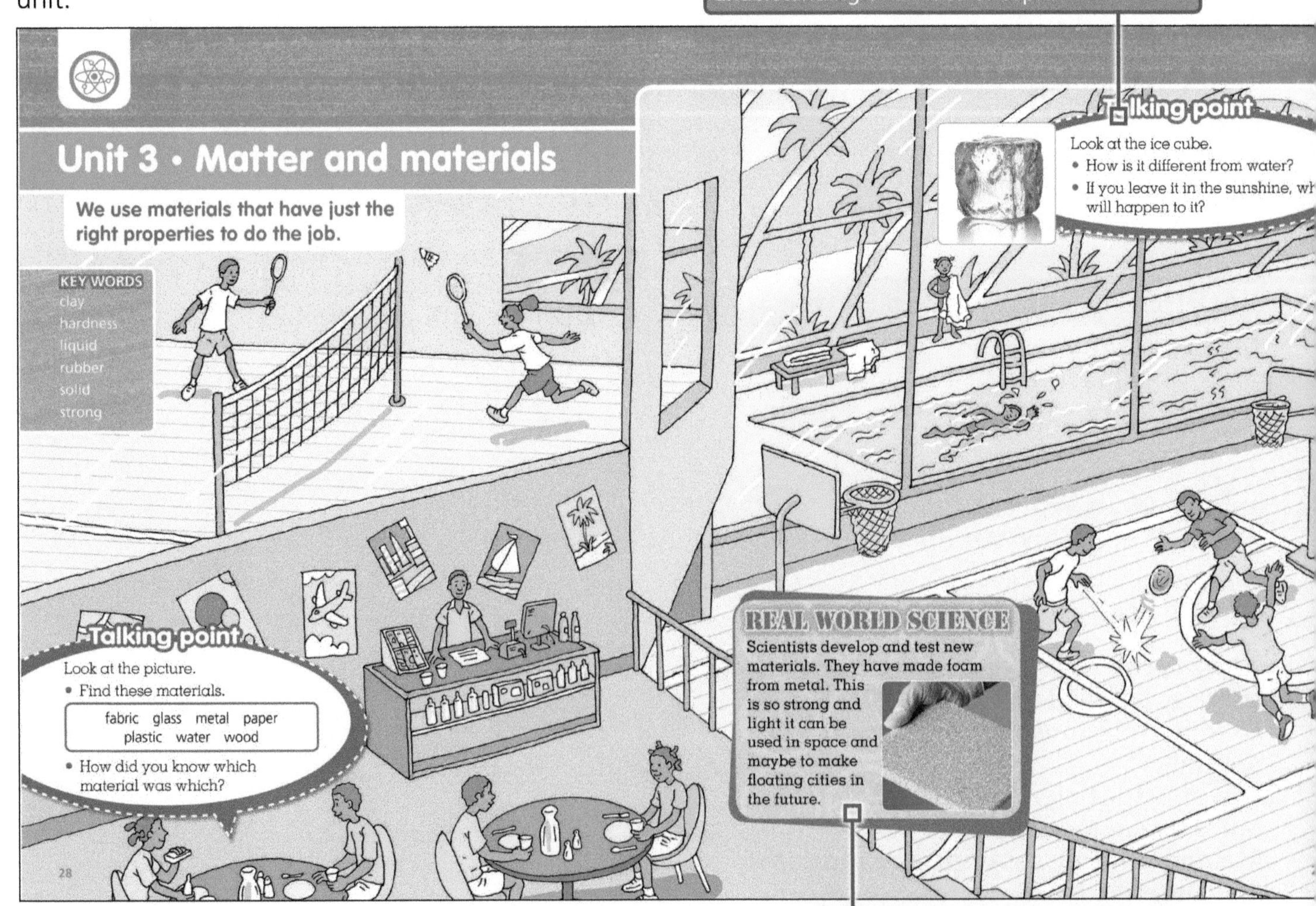

Pupils can read inspirational notes about real scientists and how science is used in the world today.

Topics and lessons

Each topic leads the pupils through a series of activities that elicit learning through doing and exploring in a series of sections. These sections allow you to work with your pupils through discussing, enquiring, researching and investigating with easily accessible materials.

Workbook page referencing indicates that a section is further supported by activities in the accompanying Workbook. These can be done in class, or as homework, to complete a section.

Materials and their uses

3 Comparing the hardness of materials

Workbook page 19–20

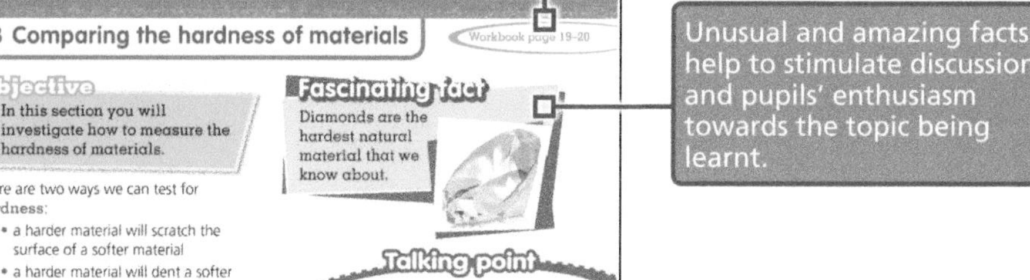

There are two ways we can test for hardness:

- a harder material will scratch the surface of a softer material
- a harder material will dent a softer material.

Talking point

How easy or hard was it to make a scratch or dent?

Finding the hardest material

1 Choose the material that you think is the hardest.
2 Choose a material that you think is not so hard.

Test 1: try to scratch the softer material with the harder material.

Test 2: tap the softer material with the harder material and try to make a dent.

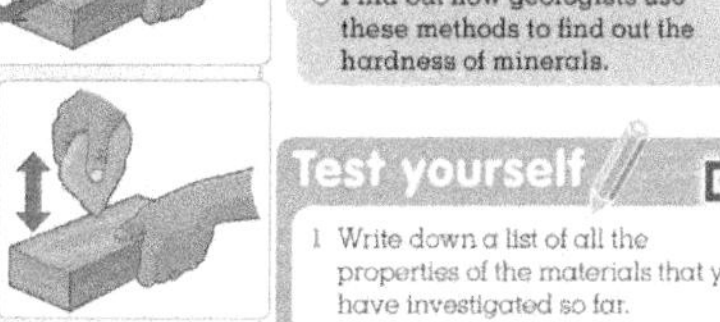

1 Rank the materials from the hardest to scratch to the easiest to scratch.

a Which material is the hardest?
b Which material is the softest?

Research task

Geologists are special scientists who study rocks and minerals.

- Find out how geologists use these methods to find out the hardness of minerals.

Test yourself

1 Write down a list of all the properties of the materials that you have investigated so far.

32 Unit 3 Matter and materials

Unusual and amazing facts help to stimulate discussion and pupils' enthusiasm towards the topic being learnt.

Scientific words and terminology are highlighted in blue throughout. Definitions can be found in the glossary at the end of the book.

These activities foster scientific investigation with particular enquiry skills, such as predicting, observing, recording, planning, evaluating and reporting.

These tasks develop pupils' research skills by encouraging them to find out more information about a topic by themselves.

At the end of most sections there are opportunities for pupils to review their own knowledge and understanding by answering short questions.

This symbol indicates that a question can be used to challenge pupils to extend their knowledge.

Safety warnings explain how budding scientists can conduct experiments safely.

Unit overview

Science projects: ideas and instructions at the end of each unit provide pupils with opportunities to create a science project to showcase what they have learnt.

Checking and testing: pupils are encouraged to review what they have learnt during the unit by completing an end-of-unit test.

You can use these tests to assess which areas pupils may need further support in.

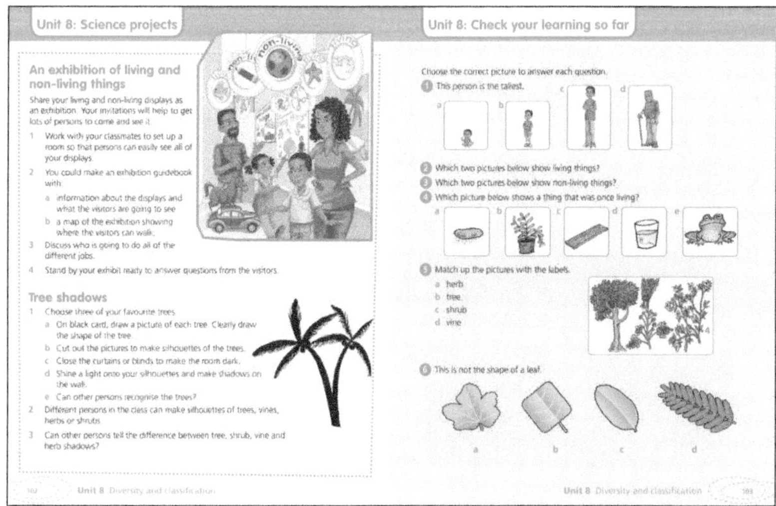

Scope and sequence

	Section	Syllabus objective(s)	Scientific enquiry
Unit 1: Structure and function			
Plants and animals develop in different ways	1 Human growth	Identify and name the different stages in the development of animals in their environment	
	2 Where do butterflies come from?		Observing and recording
	3 Frogs have a life cycle		
	4 Remember plants	Identify and name the different stages in the development of plants in their environment	Growing your own plants from seeds (part 1)
	5 How plants grow from seeds		• Growing your own plants from seeds (part 2) • Growing a seed to test if roots always grow downwards
Unit 2: Ecosystems 1			
Adaptation of organisms to their environment	1 Adapted to survive	Appreciate that organisms are adapted to their environments	
	2 How do animals and plants survive?	• Investigate how organisms adapt to their environment • Identify some features of organisms that are designed to survive in their natural environment/ecosystem	
Unit 3: Matter and materials			
Materials and their uses	1 Rubber is a material	Review different materials from Grade 1 and others such as rubber and clay	Predicting properties
	2 Materials have their own properties	List the properties of these materials	Testing the properties of materials
	3 Comparing the hardness of materials	Investigate and compare hardness and strength, etc.	Finding the hardest material
	4 Matching properties of materials to their uses	Match properties of materials to their uses	How materials are used
Designing an object	5 Designing an object for the classroom	• Design an object for a particular purpose • Share your design – drawings and plans	• Designing an object for the classroom • Presenting ideas to a group
	6 Using clay to make an object	Choose materials suitable for making objects to be used for a particular purpose	Making an object from clay
	7 Making a bouncing ball from rubber bands		Making a bouncy ball from rubber bands
	8 Making a dish for jewellery	Make an object for a particular use	Making a papier mâché dish
	9 Making a pencil case		Investigating how to make a pencil case
Solids and liquids	10 Testing solids and liquids	Describe the properties of solids, liquids and gases	Are they solids or liquids?
	11 Examples of solids and liquids	Give examples of solids, liquids and gases	
Water and ice	12 Water can be solid or liquid	Recognise that water can be solid or liquid	
	13 Changing from solid to liquid and back	Identify the conditions under which water changes from solid to liquid and back	

	Section	Syllabus objective(s)	Scientific enquiry
Liquids at home	14 Liquids are used in the home	Name examples of liquids found in the home	
	15 Measuring how liquids flow	Compare the properties of liquids used in the home	Measuring how liquids flow
Unit 4: The Earth's weather			
Measuring the weather	1 Evaporation and condensation	The importance of the water cycle	• Measuring and predicting evaporation • Observing condensation
	2 The water cycle	Draw a simple diagram to represent the water cycle	
	3 How much rain is falling?	Design and construct a simple rain gauge	Measuring the rainfall in your local area
	4 How windy is it?	Design and construct a simple anemometer to measure wind speed	• Designing and making an anemometer • Measuring the wind speed using an anemometer
	5 Weather charts	Construct a weather chart	
	6 The weather day by day	Take and compare rainfall, temperature, wind direction and wind speed on different days	
Unit 5: The Solar System			
The Sun, Earth and Moon	1 The Sun, Earth and Moon are parts of the Solar System	Name the Sun, Earth and Moon as parts of the Solar System and infer the position of the Sun at different times of the day	What happens to my shadow in the Sun?
	2 The phases of the Moon	Identify the phases of the Moon	Phases of the Moon game
	3 Day and night	Identify patterns in the occurrences of day and night	Recording day and night where you live
Unit 6: Ecosystems 2			
Feeding relationships and defence in animals	1 Eating and being eaten	• Identify feeding relationships among organisms and predator prey relationships • Simple food chains and natural defences that help animals survive	
The effects of environmental destruction	2 What is environmental destruction?	Define the terms environment and destruction and investigate factors that result in environmental destruction	
	3 Looking after the environment	Identify some ways in which environmental destruction can be prevented	
Solid waste management	4 How should we treat solid waste?	State the meaning of solid waste and identify methods of managing solid waste in the home/school/community – to include recycling to construct a toy using discarded materials	
	5 Cleaning up litter	Litter in schools and clean-up projects	My clean-up project
Making different sounds and musical instruments	6 The sound of music	Name the human organ that is stimulated by sound and classify sounds by pitch and loudness	Investigating pitch
	7 Making a musical instrument	Construct and use simple musical instruments	Making a musical instrument

	Section	Syllabus objective(s)	Scientific enquiry
Unit 7: Forces, motion and structures			
Effects of forces	1 Using forces in sports	Identify forces used to create movement or change in given situations	Using forces to make objects move
	2 Changing direction	Demonstrate ways in which motion can be changed	• Predicting what happens in a collision • Changing the direction of a moving object • Directing a toy car using only pushes and pulls
Simple mechanical devices	3 Mechanical devices around us	Identify simple mechanical devices	• Mechanical devices in the classroom • Why are nails the shape they are?
Unit 8: Diversity and classification			
Human variation	1 The same and different	State ways that people are alike and are different	
	2 How are we the same and different?	Group themselves according to similarities	A class comparison
Living and non-living things	3 Alive or not alive?	Examples of living and non-living things	
	4 Living and non-living display	Make a presentation displaying living and non-living things	
Characteristics of living things	5 Moving and growing	Characteristics of living things: moving and growing	How do plants move?
	6 Feeding and reproducing	Characteristics of living things: feeding and reproduction	
How do plants differ?	7 Plants are not all of one type	Define plant, tree, shrub, vine and herb	
	8 Different plants	Identify different kinds of plants	Finding different types of plants
	9 Leaves are important	List different uses of leaves and name two types of leaves found in plants	Collecting and sorting leaves

	Section	Syllabus objective(s)	Scientific enquiry
Unit 9: Energy			
Making use of devices	1 Devices that use energy	• List devices in the home and community that use electricity or other forms of energy • State how these devices have improved life and what difficulties may be encountered without them	
	2 Old and new devices	• Appreciate that people use energy to solve some of their problems • Compare old and new technologies and infer that people keep inventing new things	
Windmills and waterwheels	3 Wind and water are sources of energy	Infer that wind and water are sources of energy and identify and observe devices that use these energy sources	How big do the sails of a windmill need to be?
Unit 10: Earth's resources			
Dealing with pollution	1 The dangers of litter	Review the dangers of litter	Finding out how your school manages litter
	2 Litter in schools is a problem	Discuss how the problem of litter in schools can be avoided	Encouraging the use of litter bins
	3 Organising and taking part in a clean-up campaign	Organise and participate in a clean-up project	Organising a clean-up campaign
	4 Air contains pollutants	Identify at least two air pollutants found in a particular area	Investigating pollutants in the air
	5 Pollutants in the air affect people	Discuss how pollutants affect people's activities	
	6 Collecting and removing dust from the air	Construct a trap for collecting dust from air	Making an air trap to filter dust from the air
	7 Collecting dust with a dust filter	Compare the amount of pollution found in different areas using the constructed air trap	Making a dust filter
	8 Protection from air pollution	Identify and compare the devices developed to protect workers from air pollution	

Unit 1 • Structure and function

Living things grow and develop to make the millions of different living things on Earth.

KEY WORDS
adult
develop
egg
grow
larva
pupa
seed

Talking point

- Does a baby bird look like an adult bird?
- Do all animals hatch from eggs?

Talking point

- Do butterflies come from eggs?
- Do baby frogs look like adult frogs?

Talking point

- How long does it take a coconut tree to grow?
- Why do living things start small and then grow?

Fascinating fact

Your ears and nose will grow all of your life. Your eyes are the same size they were when you were born.

REAL WORLD SCIENCE

Scientists study living things to learn about how they grow. Here they are measuring, tagging and tracking the animals.

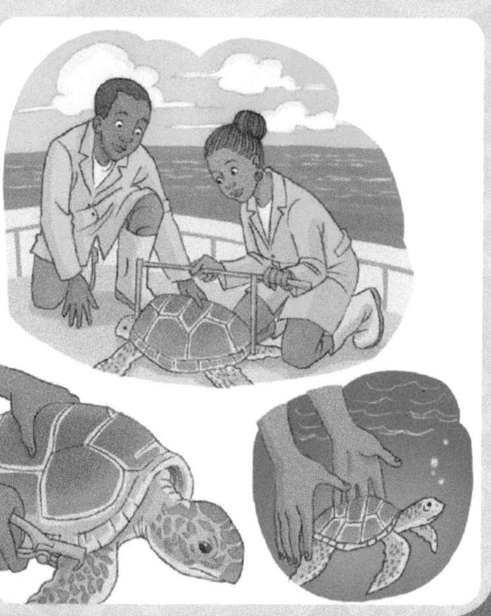

1 Human growth

Workbook page

Objective

In this section you will learn the different stages as humans grow and change.

Talking point

Do people grow all of their lives?

Your body is made up of different parts.

1. Show your partner where these parts of the body are. Take it in turns to point to them.

head arm leg shoulder knee
nose eye ear hand finger

Our bodies change as we get older.

Some things change as we **grow** and **develop**. Some things stay the same. We have the same eye colour all through our lives.

a middle-aged **adult**

f child

d older adult

2 years old, 15 years old, 45 years old

3 months old, 5 years old, 25 years old, 75 years old

Talking point

- What stays the same all the way through a person's life?
- What changes as a person gets older?

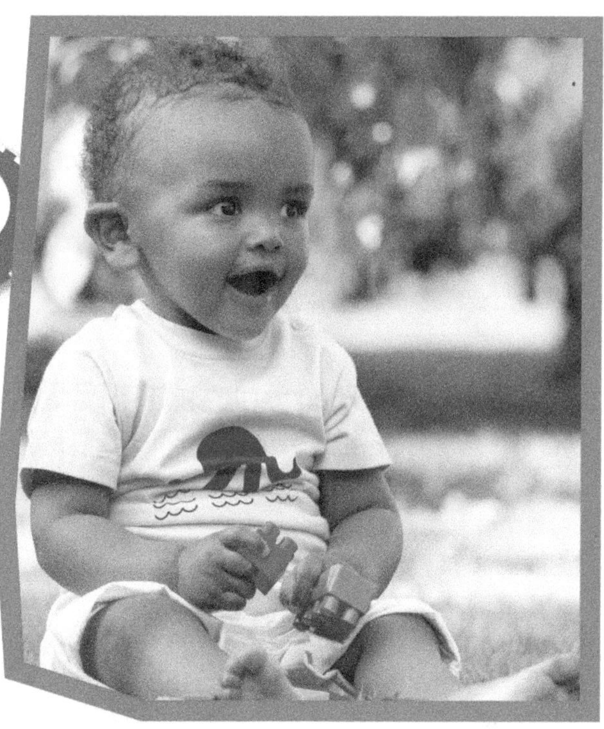

2 Match the pictures to the ages in the timeline.

We cannot see all of the changes.
Our voices change as we get older.

3 a Make the sounds a baby makes.
b Talk in your own child's voice.
c Now try to talk like an adult.

c infant

b adolescent

e young adult

g baby

Test yourself

1 Which statements are true and which are false?
   a A baby can run and jump.
   b We have the same hair colour all our life.
   c We have the same eye colour all our life.
   d We keep growing taller all our life.

2 Where do butterflies come from?

Workbook page

Objective

In this section you will learn the different stages in the life cycle of a butterfly.

Many young animals, such as humans, dogs and goats, look like their parents. They are just smaller.

This young animal can look after itself.

Many young animals cannot look after themselves.

1 Which is the caterpillar's parent?

a

b

c

Talking point

How is the young goat being looked after?

Life cycles

A life cycle shows all of the changes during an animal's life. Look at the picture of the life cycle of a butterfly.

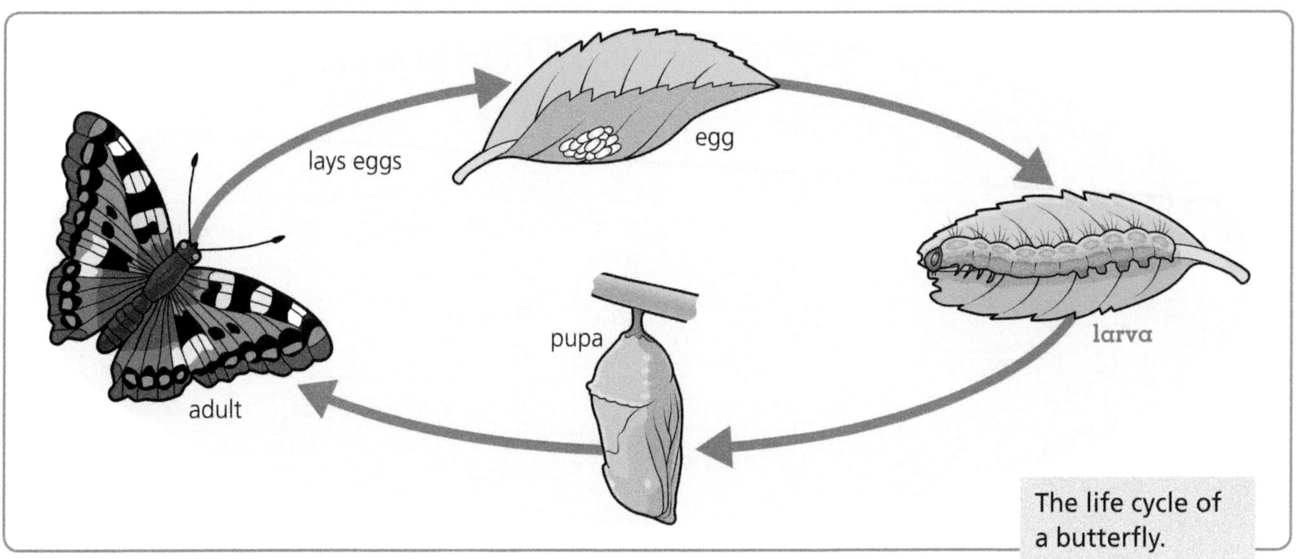

The life cycle of a butterfly.

The butterfly looks different at different stages of its life.

The **pupa** is when the caterpillar is changing to a butterfly.

2 a What is the caterpillar stage called?
b What can the butterfly do that the caterpillar cannot do?

Talking point

- Why do butterflies have a caterpillar stage?
- Why not be a butterfly all of its life?

Fascinating fact

Some butterflies only lay their eggs on one type of plant. The caterpillar will then eat just this plant.

Observing and recording

Look for caterpillars and adult butterflies.

1 How many different types of each can you find?
2 Look up the names of the butterflies you have found and record them.

Test yourself

1 Here are the stages in the life cycle of a butterfly. Put them in the correct order.

butterfly egg larva pupa

3 Frogs have a life cycle

Workbook page

Objective

In this section you will learn the different stages in the life cycle of a frog.

There are many different types of frogs. Some are bright colours. Other frogs are green. All frogs have the same life cycle.

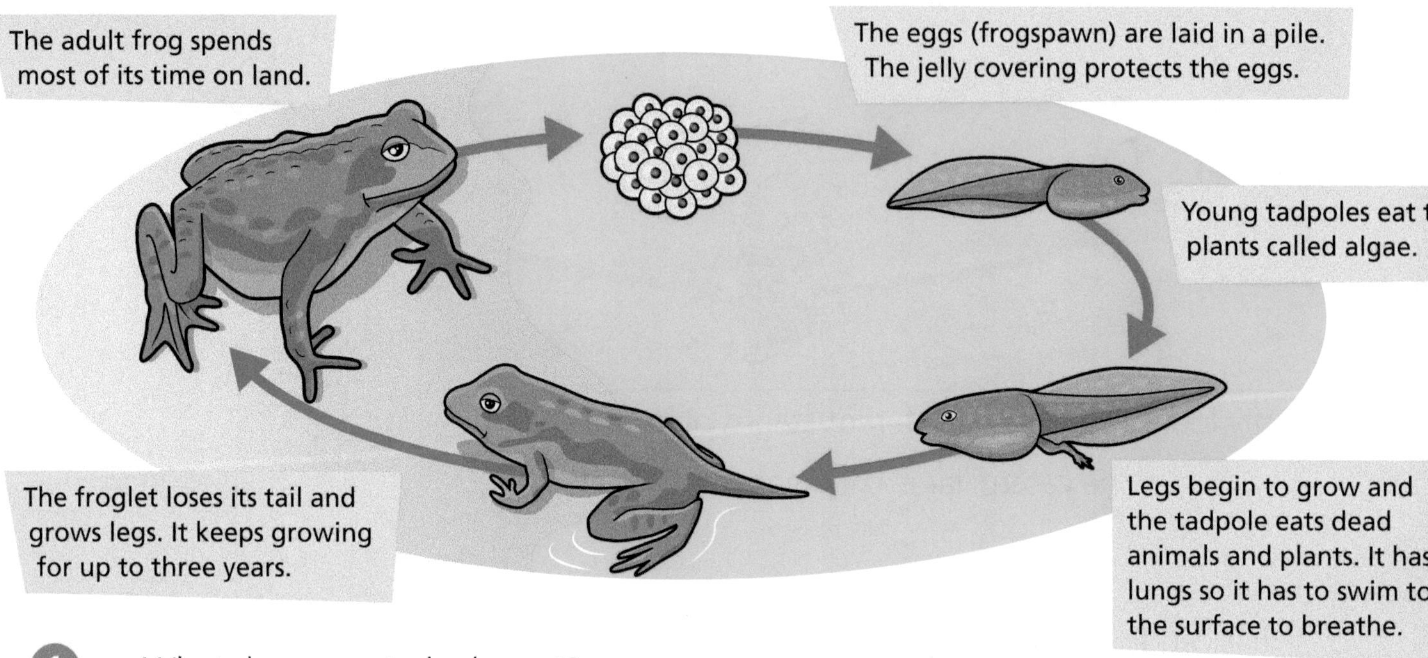

The adult frog spends most of its time on land.
The eggs (frogspawn) are laid in a pile. The jelly covering protects the eggs.
Young tadpoles eat t plants called algae.
Legs begin to grow and the tadpole eats dead animals and plants. It has lungs so it has to swim to the surface to breathe.
The froglet loses its tail and grows legs. It keeps growing for up to three years.

1 a What do young tadpoles eat?
  b What is a young frog called?

Research task

- Find out how long it takes for frog eggs to hatch into tadpoles.

Fascinating fact

Some animals shed their skin as they grow. Snakes and lobsters do this. Their skin isn't stretchy so they get rid of it and grow another one.

Test yourself

1 Here are the stages in the life cycle of a frog. Put them in the correct order.

adult frog   tadpole with legs
egg   froglet   young tadpole

4 Remember plants

Workbook page 8

Objective

In this section you will remember the important parts of flowering plants.

Plants are living things. They need food and water.

1 Talk about the pictures of the plant and the flower with your partner.

a Point to and name all the parts of the plant.

b Point to and name the parts of the flower.

flower
leaf
ovary
petal
roots
sepal
stamen
stem

Growing your own plants from seeds (part 1)

1 a Plant five seeds in pots.
b Observe your plants as they grow.
c Look for any changes.

2 What measurements will you take?

3 Write down two things your plants will need.

Talking point

Think about the plants you can see growing near your school.

- What changes can you see as the plants grow?
- Do any of the plants make flowers?

Test yourself

1 Which part of the plant attracts insects?

2 Which part of the plant holds the plant in the soil?

5 How plants grow from seeds

Workbook pages 9

Objective

In this section you will learn the different stages in the life cycle of plants.

Look at the stages in the life cycle of a flowering plant.

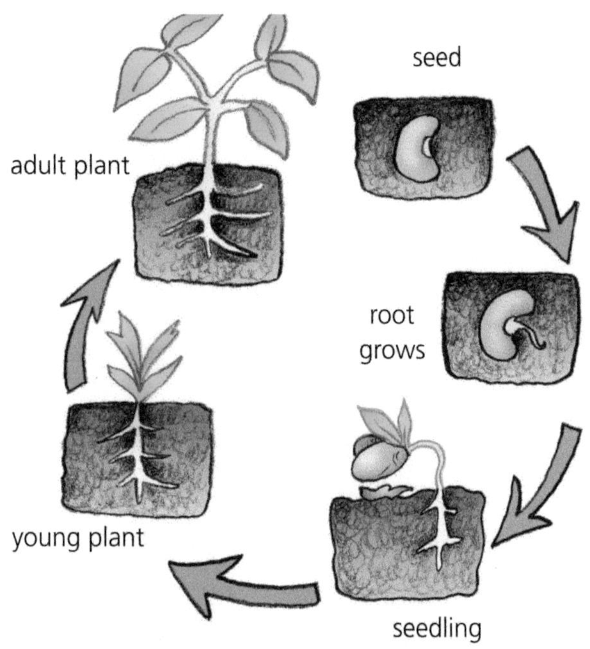

Growing your own plants from seeds (part 2)

1 Observe and measure your plants as they grow. Record your results.
2 Are your plants following the stages shown in the picture?

  Do roots always grow downwards?

Remember

A good scientist does not believe things without testing!

REAL WORLD SCIENCE

Scientists can use powerful microscopes to see inside the cells of plants. This helps them to understand how plants grow.

Growing a seed to test if roots always grow downwards

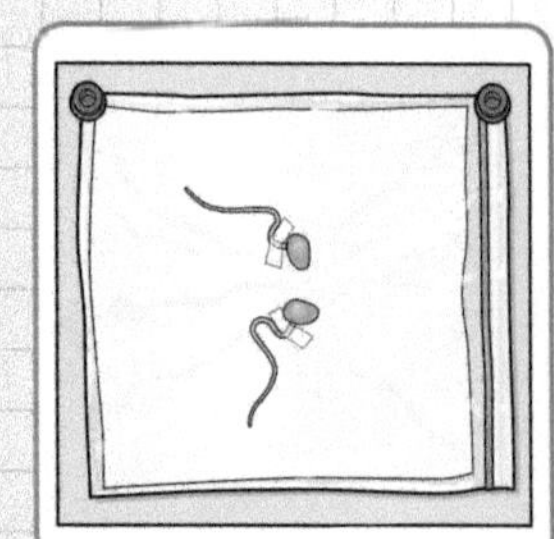

1 Seal two damp bean seeds in a plastic bag. Stretch the bag tightly and fix it to some cardboard.
2 Stand the cardboard against a wall. Every few days, turn the cardboard a quarter turn.
3 Observe what happens.
4 Draw a diagram to show what has happened to the roots.

Research task

- Do all plants have the same life cycle?

From seeds to flowers

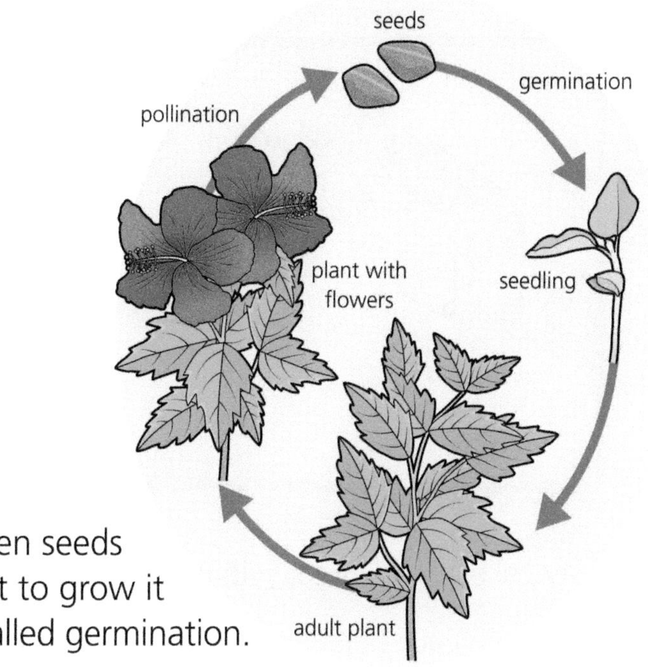

When seeds start to grow it is called germination.

Flowers need pollen from another flower to make seeds. This is called pollination.

2 The seeds of some plants are inside juicy parts that are good to eat. What are these juicy parts called?

fruits leaves roots stems

3 Some seeds start to grow. What is this called?

breathing germination
movement pollination

Fruit trees have almost the same life cycle as other flowering plants, but they produce fruit with the seeds inside.

Some trees produce cones with seeds inside.

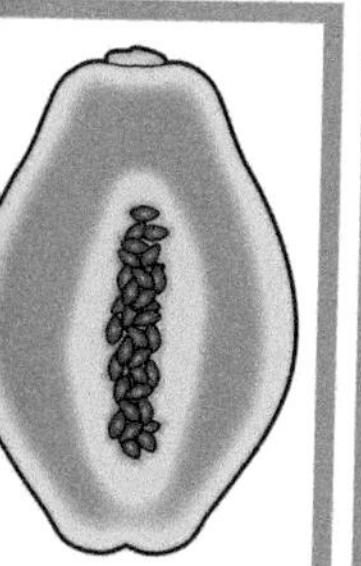
t with seeds.

Cone with seeds.

Fascinating fact

The tallest trees in the world grow from tiny seeds. Giant redwoods can grow to 100 metres high. Their seeds are 5 mm long!

Check the length of this seed with a ruler.

Research task

- Find out which plants have the largest seeds.
- Do these seeds grow into tall plants?

Test yourself

1 Copy out the passage and write in the missing words.

You will need some words more than once.

Flowering plants grow from ☐. In the soil these ☐ to make ☐. These ☐ then grow into adult plants. The plants then develop ☐ and these make more ☐. Some ☐ are found inside juicy ☐.

flowers seeds fruits
germinate seedlings

Presenting the life cycle of a butterfly

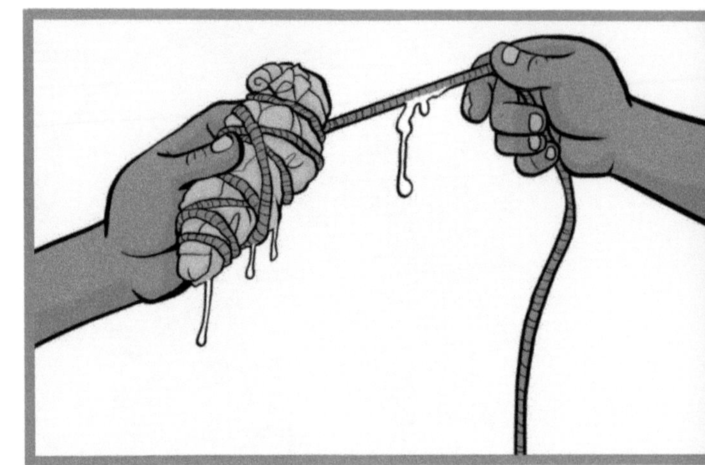

1 Make models to show the life cycle of a butterfly.
   a Make small eggs using modelling clay.
   b Draw and colour a butterfly and a caterpillar on paper or card. Cut them out.
   c To make the pupa stage, dip some string or thread into glue. Wrap the string around a small piece of scrunched up paper. Let it dry. You have made a pupa.
2 Use your models and drawings to make a display called 'The Life Cycle of a Butterfly'.
3 Present your life cycle. Leave out one stage of the life cycle. Ask your class to name the missing stage.

Growing different plants from seeds

1 Plant some seeds in a pot of moist soil or compost.
2 Take a photograph every few days to show each stage.
3 Make a poster about growing plants for your science fair.

Remember
- Label the parts of the plant.
- Label any stages that you see.

Unit 1: Check your learning so far

What can you see in each picture?

1 Name the stage in the life of a human.

a

A adolescent
B adult
C baby
D infant

b

A adolescent
B adult
C baby
D infant

2 Name the stage in the life cycle of a butterfly.

a

A adult
B egg
C larva
D pupa

b

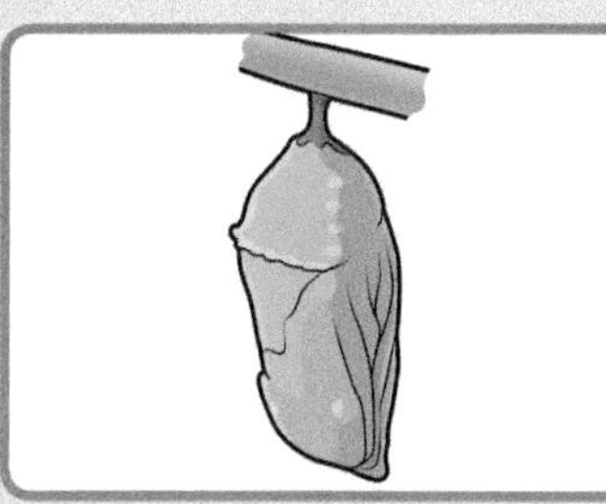

A adult
B egg
C larva
D pupa

3 Read the definitions. Match each definition to the correct word from the word box. You will not need all the words.

a The first stage in the life cycle of a frog.
b The part of a plant where seeds are made.
c This happens when the seed is just starting to grow.
d The name of the stage when the plant has small roots and shoots.

adult egg flower germination leaf pollination seed seedling

Unit 2 • Ecosystems 1

Living things are adapted to their environment so they fit in and survive.

KEY WORDS
adapted
habitat
physical characteristic
survive

Talking point

Look at the animals in their coral reef habitats. Which animals:

- are good at hiding?
- are looking for food?
- would be easy to eat?
- would be difficult to eat?

Remember

A habitat provides shelter, food and a safe space for animals.

Fascinating fact

Corals are not plants. They are animals that are related to jellyfish!

REAL WORLD SCIENCE

Scientists are looking for new medicines by testing chemicals made by animals and plants in coral reefs.

Talking point

- How many different animals can you see?
- Why are some animals more difficult to find than others?

Talking point

- Why do frogs not live in nests?
- Why does the mongoose not live under water?

Fascinating fact

Woodlands and forests are almost the same, but not quite. If you stand in woodland and look up you can mainly see the sky. In a forest you can mainly see trees!

1 Adapted to survive

Workbook pages 1

Objective

In this section you will:

- learn that animals and plants are adapted to where they live
- identify some adaptations that help animals and plants to survive.

Fascinating fact

Animals and plants are **adapted** to **survive** in their **habitats**. This means that they live best in a particular place.

What you see when you look at an animal is called its **physical characteristics**.

1 Here are four different physical characteristics of animals.

feathers	fur	moist skin	scales

a Write the physical characteristics in your notebook.

b Think of animals you know of that have these characteristics. Write a list.

2 Look at the pictures of animals.

a Which animals do you think are adapted to live in dry habitats?

b Which animals are adapted to live in wet habitats?

The animals and the plant in the pictures have many characteristics that help them to survive.

3 Which picture shows:

a a plant adapted to stop it falling over?

b an animal adapted to wade in water?

c an animal adapted so it can swim well

d an animal adapted to fly?

Test yourself

1 What is one way that birds are adapted to their habitat?

2 What is one way that fish are adapted to live in water?

2 How do animals and plants survive?

Workbook pages 13–16

In this section you will learn about how animals and plants are adapted in order to survive.

Thick fur here to protect me from the harsh sunshine.

Narrow nostrils and two sets of eyelashes to keep the sand out.

.arge, flat feet to stop ne sinking in the sand.

Thin fur here to let me lose heat.

Short, sharp beak to catch fish.

Closely packed feathers to keep me waterproof and streamlined.

Short wings to help me to swim, but I cannot fly.

Lots of fat to keep me warm.

Short, stiff tail to help me stand up on land.

Adaptations help animals and plants to survive and reproduce in the habitat where they live. The adaptations make it very difficult for them to live in other habitats.

A little egret.

a In your notebook, draw the little egret.

b Now draw a circle around any adaptations you can see that help this bird to live near water and catch fish.

Test yourself

Choose the correct words.

1 Which two adaptations help birds to fly?

big eyes   light weight   long legs   wings

2 Which three adaptations help fish to swim?

bright colours   fins   smooth scales   streamlined shape

Designing an animal

1 Design an animal that lives on a planet that is cold and covered in water. The animal eats large fish and burrows under the mud at night.

2 Draw your design for your animal and label the adaptations.

3 Make a model of your animal and place it in its habitat. You could make a cardboard box into part of the planet.

Building a mangrove tree

1 Use straws and tape to make a model of a mangrove tree.

2 Stand it in loose sand and try to push it over.

  a Was it easy or difficult? Why?

  b How do the roots help to keep it stable?

3 Now make a model of a tall, thin tree – a pine tree or a palm tree.

4 Stand it in loose sand and try to push it over. Was it easy or difficult? Why?

5 Present your models to someone. Explain how the mangrove tree is adapted to its habitat.

Unit 2: Check your learning so far

How are these animals and plants adapted to their habitats?

1. Find the correct animal or plant for each description.

   a This animal is adapted to live in water.
   b This animal is adapted to live in cold places.
   c This animal has wings so it can fly.
   d This animal can burrow underground to keep cool.
   e This plant is adapted to live in water.
   f This plant is adapted to live in loose sand and mud.
   g This plant is adapted to live in the desert.

desert iguana

seaweed

mangrove tree

cactus

polar bear

swallow

shark

2. Now write down one adaptation that helps each animal or plant to live in its habitat.

Unit 3 • Matter and materials

We use materials that have just the right properties to do the job.

KEY WORDS
clay
hardness
liquid
rubber
solid
strong

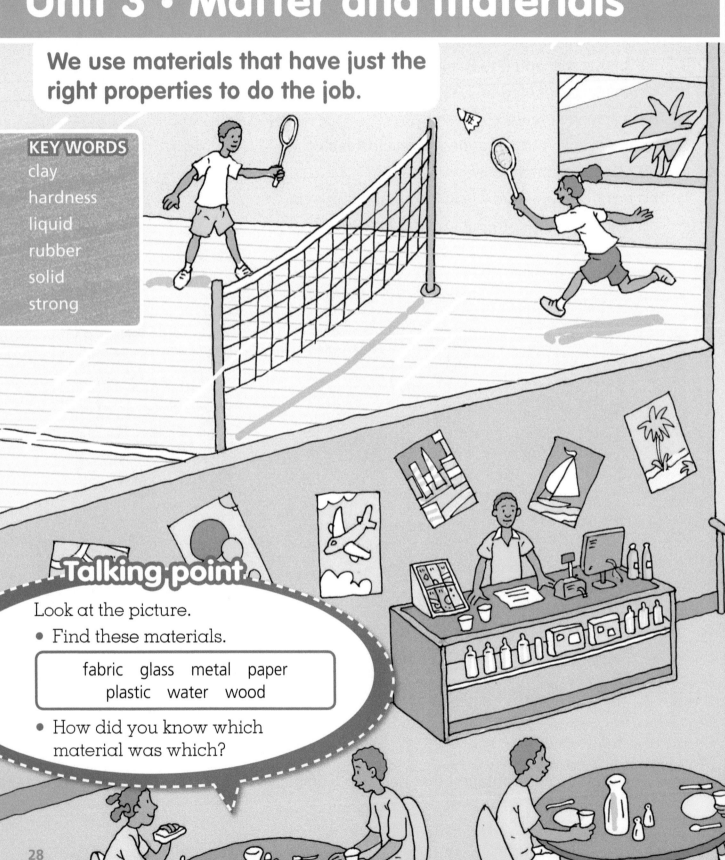

Talking point

Look at the picture.

- Find these materials.

fabric glass metal paper
plastic water wood

- How did you know which material was which?

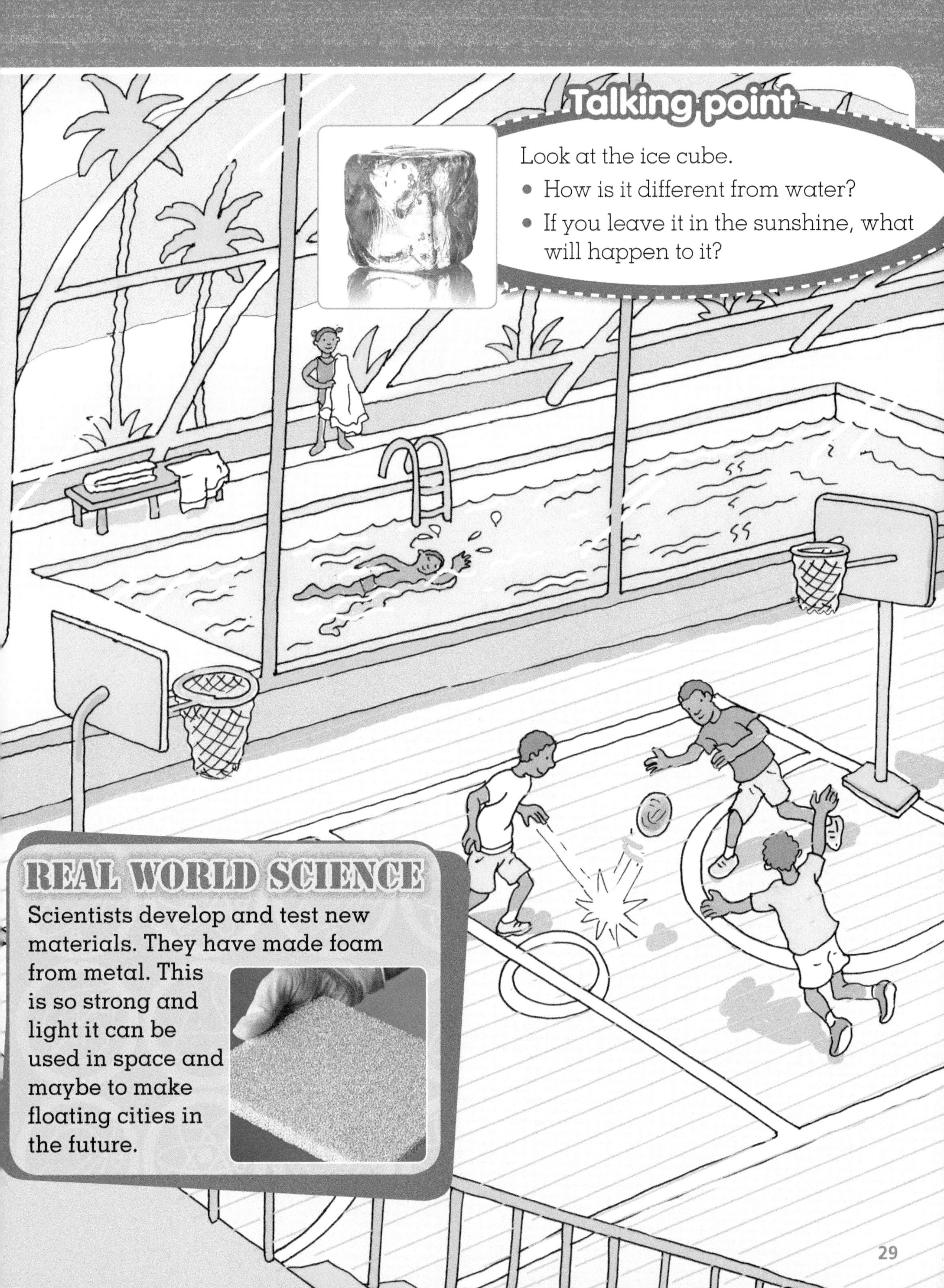

Talking point

Look at the ice cube.

- How is it different from water?
- If you leave it in the sunshine, what will happen to it?

REAL WORLD SCIENCE

Scientists develop and test new materials. They have made foam from metal. This is so strong and light it can be used in space and maybe to make floating cities in the future.

Materials and their uses

1 Rubber is a material

Workbook page

Objective

In this section you will learn more about materials.

You have already learnt about glass, paper, wood, plastic and metals.

1 Look at these objects made of **rubber**. How is rubber being used?

Fascinating fact

Natural rubber comes from a tree called *Hevea brasiliensis*. The picture shows how natural rubber is collected.

Collecting rubber is slower and more expensive than using plastics, which are human made. Most of the rubber used today is not natural.

Predicting properties

- What do you predict the properties of rubber will be?

Talking point

Have you seen rubber being used?

Research task

- Find out how rubber was used in the past.
- What are some modern uses of rubber? (Use the pictures on this page to give you some clues.)

2 Materials have their own properties

Workbook page 18

Objective

In this section you will explore the properties of different materials.

The properties of a material describe how it acts and what it can do.

What properties of materials did you test for in Grade 1?

Fascinating fact

Clay is a material that comes from rock or soil. When clay is heated and becomes hard we call it pottery.

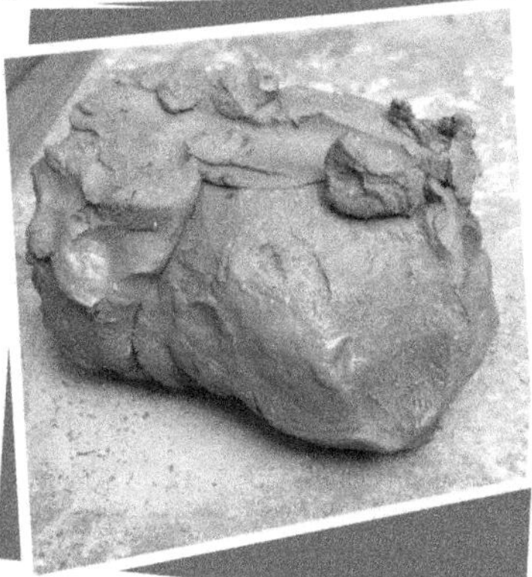

Talking point

Look at the pictures.

- Where does clay come from?
- Is it difficult to think that the clay could make an attractive pot?

Testing the properties of materials

1 Test the materials your teacher has given you to find out if it is:

- stretchy – pull the material
- bendy – try to fold the material
- shiny – look at the material in light
- hard – feel the material and compare it to soft materials
- 'clangy' – tap the material: does it make a sound?
- **strong** – can the material hold a book or will it break?

2 Draw a table in your notebook to record your results.

3 Write down all the properties of the materials you have tested.

Test yourself

1 Name a material that makes a 'clangy' sound.

2 Name a material that is stretchy.

3 Comparing the hardness of materials

Workbook pages 19

Objective

In this section you will investigate how to measure the hardness of materials.

There are two ways we can test for **hardness**:

- a harder material will scratch the surface of a softer material
- a harder material will dent a softer material.

Finding the hardest material

1 Choose the material that you think is the hardest.

2 Choose a material that you think is not so hard.

Test 1: try to scratch the softer material with the harder material.

Test 2: tap the softer material with the harder material and try to make a dent.

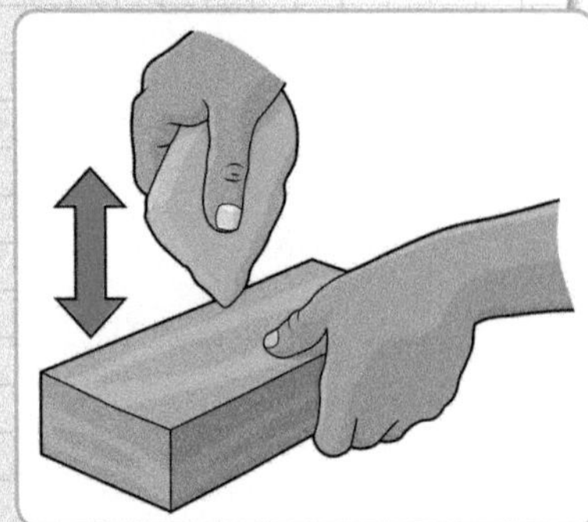

Fascinating fact

Diamonds are the hardest natural material that we know about.

Talking point

How easy or hard was it to make a scratch or dent?

1 Rank the materials from the hardest to scratch to the easiest to scratch.

a Which material is the hardest?

b Which material is the softest?

Research task

Geologists are special scientists who study rocks and minerals.

- Find out how geologists use these methods to find out the hardness of minerals.

Test yourself

1 Write down a list of all the properties of the materials that you have investigated so far.

4 Matching properties of materials to their uses

Workbook page 21

Objective

In this section you will investigate the properties of materials and how they are used.

People choose materials to make different objects because of the properties of the materials.

glass	metal	paper	plastic	pottery	rubber	wood

1 a Look at the names of the materials. In your notebook, write down a use for each material.

b Draw a picture of an object made from each material.

How materials are used

1 Look at the list of materials at the top of the page.
2 Look at different objects around your classroom. Find one object made from each material.

Fascinating fact

Gold is a shiny metal. This is one of the properties that makes it useful for making jewellery.

2 a What other materials are used to make jewellery?

b Do they have the same properties as the metals you have investigated?

Talking point

- Look at the materials that other pupils have found.
- Did you all identify the materials correctly?

5 Designing an object for the classroom

Objective

In this section you will:
- design an object
- share your ideas with your class.

The properties of a material make it good for some purposes but not good for other purposes.

Fascinating fact

People used to use bags because pockets had not been invented.

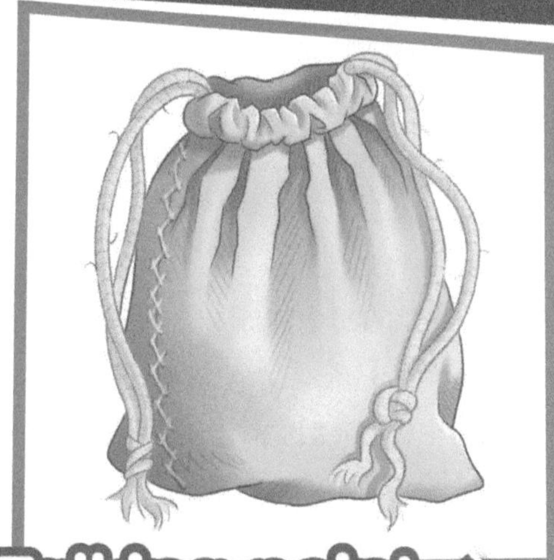

Talking point

What could you carry in this bag?

1. a What material is the bag made from?
   b Do you think the bag is strong?
   c Can it hold much weight?

Designing an object for the classroom

Work in a small group. You will need to work as a team and communicate with each other.

1 Choose one of these objects:
- an object to store crayons
- an object to store counters or dice
- an object to store paper
- an object to store toys.

2 Start your design.
- How big will the object be?
- What materials will you use?
- Will you sew or glue the materials together?

3 Draw your design. Label the materials you use in your design clearly.

Workbook page 22

Sharing your design drawings and plans

Fascinating fact

People design objects as a job. Every modern device you see has been designed by people.

REAL WORLD SCIENCE

Scientists have developed many materials that are used in the design of modern devices such as laptops and smartphones.

Talking point

- Who will show the design to the class?
- Will you all talk about the design?

Presenting ideas to a group

1 Decide who will talk about the different parts of your design.
2 Practise your talk.
3 Display your design to the class. Try not to stand directly in front of your design, so that everyone can see it.
4 Tell the class what the object is for.
5 Point out all the things in your design.
6 Talk about the properties of the materials you have chosen.

Remember

- Talk slowly and clearly.
- Use a ruler to point at things in your design.

6 Using clay to make an object

Workbook page

Objective

In this section you will make a pot from modelling clay.

Talking point

- Why is the clay fired in a kiln?
- What does this do to the clay?

Clay is used to make pottery and ceramics. Clay is a heavy, sticky type of soil. It is made from tiny bits of minerals mixed with water.

Not all clay has to be fired in a kiln.

Making an object from clay

1 Look carefully at the clay. Can you see tiny pieces of sand and rock?
2 Squeeze and squash the clay.
3 Decide what object you want to make from the clay.
4 Use your hands to mould the clay.
5 Place the object in a warm, dry place.

When it is dry you can decorate it with paint.

Fascinating fact

To make pottery, clay is fired (baked) in a very hot oven. The oven is called a kiln.

1 What properties of clay make it good for the object you chose?

Research task

- What other objects can you make from clay?

Test yourself

1 What is clay made from?

7 Making a bouncing ball from rubber bands

Objective

In this section you will make a bouncing ball from rubber bands.

In your notebook, write down all the properties of rubber.

Fascinating fact

Golf balls used to be made out of rubber bands.

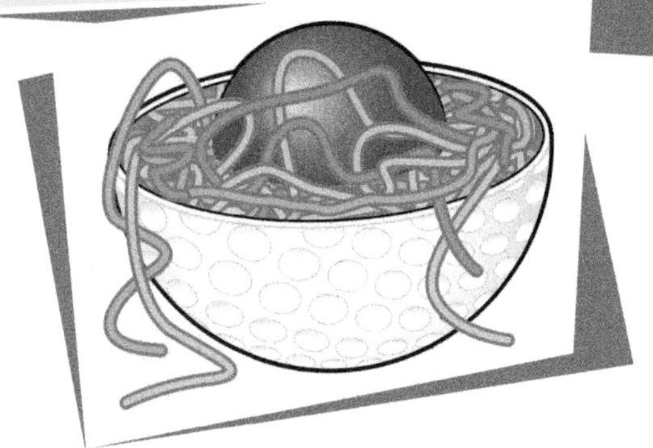

1 a Which property of rubber makes it a good material for a ball?

b How can you use rubber bands to make your own bouncy ball?

Research task

○ How are golf balls made today?

Making a bouncy ball from rubber bands

1 Squash up a rubber band.

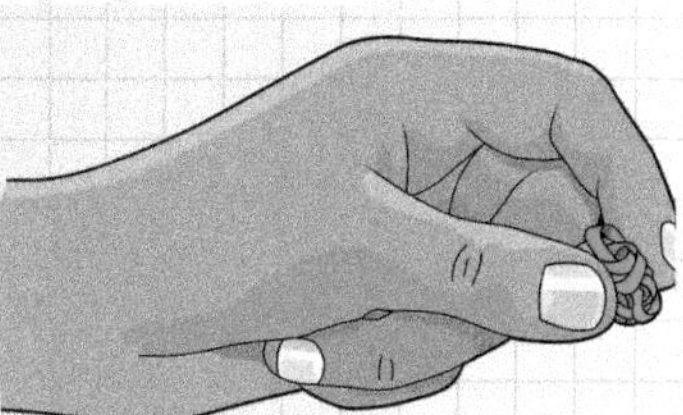

2 Wrap another rubber band around it. Ask a partner to help you.

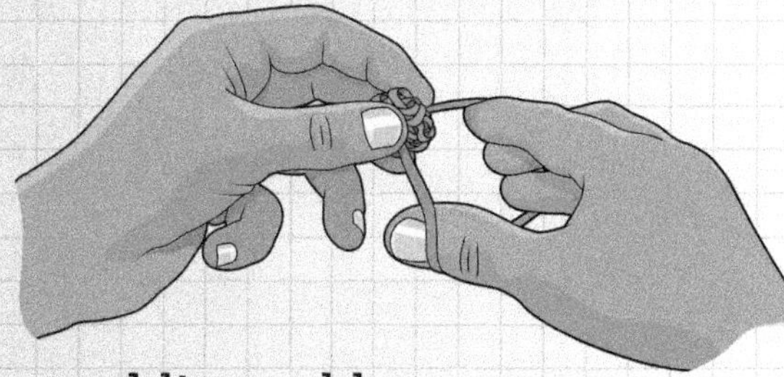

3 Keep adding rubber bands to make a ball.

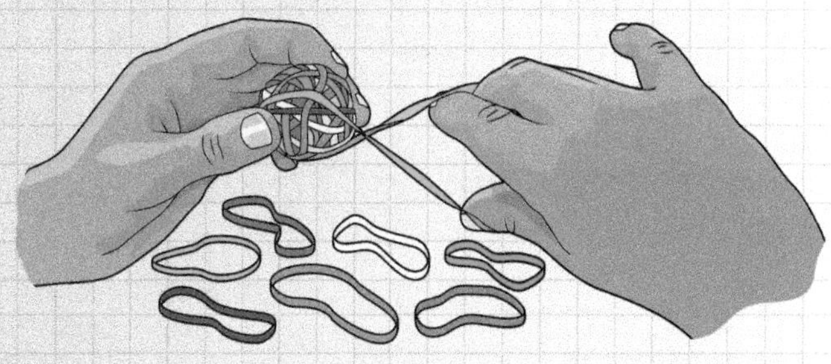

4 Try to bounce the ball.

a Does it bounce?

b How high can you make it bounce?

c Compare your ball to your classmates' balls.

2 Now design your own object made from rubber bands.

8 Making a dish for jewellery

Workbook page

Objective

In this section you will make an object to store jewellery.

Making a papier mâché dish

You are going to make a dish for storing jewellery out of papier mâché.

1 Soak strips of newspaper in flour and water paste.

2 Build up layers of paper strips on a balloon.

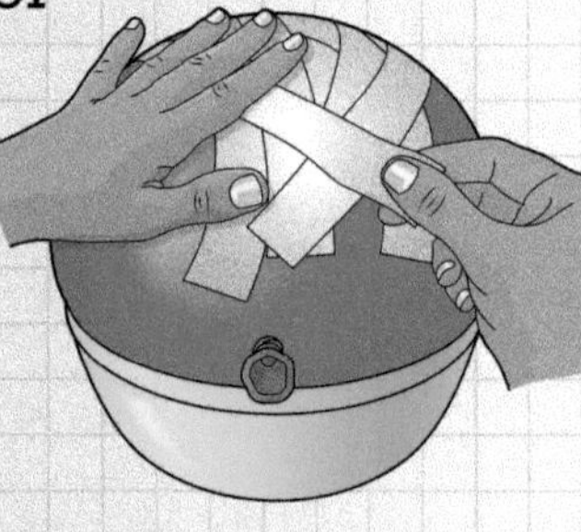

3 Leave the papier mâché to dry.
4 Use a pin to burst the balloon.
5 Use scissors to trim the edges, then paint your dish.

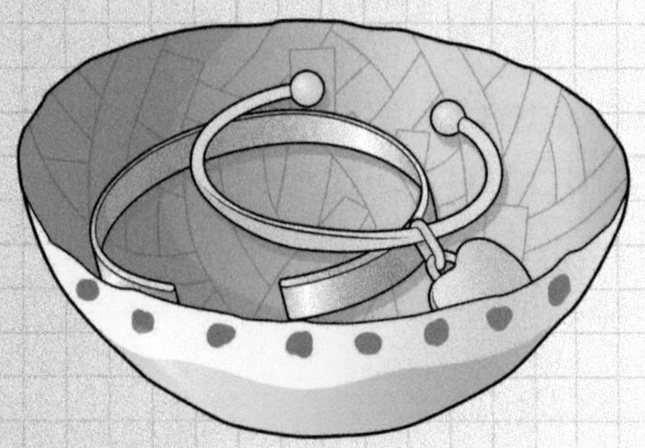

Fascinating fact

Papier mâché can be used to make masks for the theatre. Maybe you can make a mask.

1 a What are the properties of papier mâché?
b Why is it a good material for making a dish?

Research task

- In which country was papier mâché invented?

Test yourself

1 Name the three things used to make papier mâché.
2 Name another material that could be used to make a dish.

9 Making a pencil case

Objective

In this section you will choose a material and use it to make an object.

Fascinating fact

Pencils were invented over 300 years ago. In old pencils, the dark graphite in the middle was wrapped with string not wood.

Materials can be soft, bendy, stretchy and strong. They can be hard. Sometimes they can be transparent. When a person designs an object they use materials that have the right properties for that object.

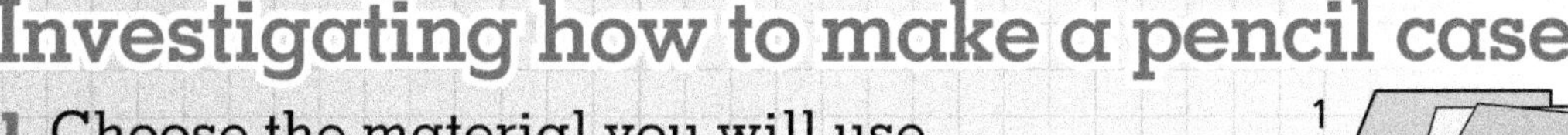

Investigating how to make a pencil case

1 Choose the material you will use.

2 Follow the steps.
   a Make sure your pencil case will be big enough to hold your pencils.
   b Help your partner, but each make your own pencil case.
   c You can sew or glue the sections together. Ask your teacher to help with the gluing.

3 Review your design.
   a Is the material strong enough to hold your pencils?
   b Is the material easy to carry to and from school?
   c What will happen to the material if it gets wet?

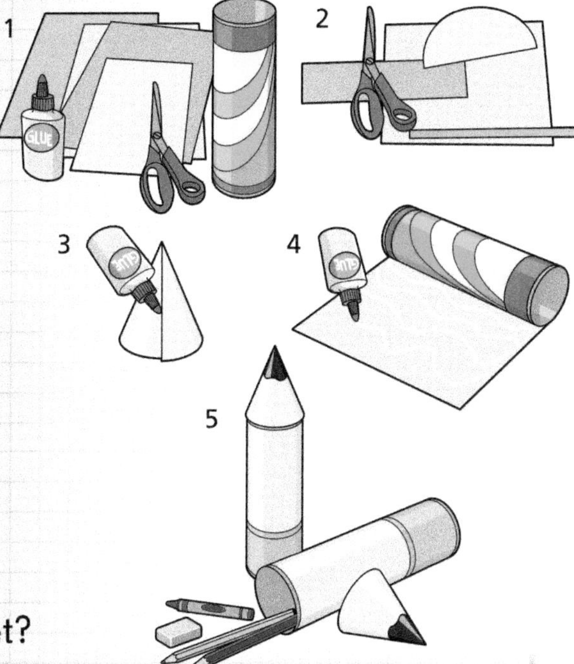

Talking point

- Have you got a pencil case?
- What material is it made from?
- Why do you think the designers of your pencil case chose this material?

10 Testing solids and liquids

Workbook page

Objective

In this section you will explore the properties of solids and liquids.

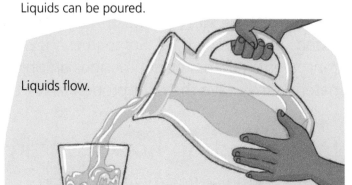

Solids keep their shape.
Solids are usually hard.

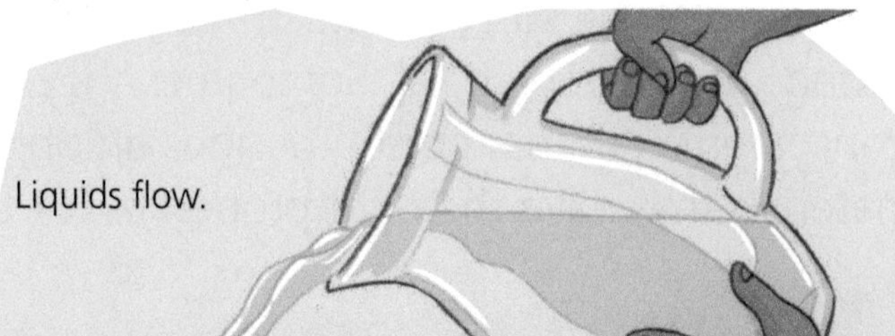

Liquids can be poured.
Liquids flow.
Liquids change shape depending on the container they are in.

Remember

To test if a material is hard you try to scratch or dent it with another material.

Are they solids or liquids?

1 Look at the materials that your teacher has given you to investigate. Predict if each material is a **solid** or a **liquid**.

2 Test the materials to check your predictions. Ask these questions about each material.
- Is it hard?
- Does it keep its shape?
- Can you pour it?
- Does it flow?
- Does it change shape in different containers?

3 Decide whether each material is a solid or a liquid.

4 Did you make the correct predictions?

Test yourself

1 Which one of these is a property of liquids?

hard   yellow   keeps its shape
takes the shape of its container

11 Examples of solids and liquids

Workbook page 26

Objective

In this section you will name some samples of solids and liquids.

The properties of a solid are that it keeps its shape and cannot be poured.

Talking point

The picture shows someone pouring sugar.

- Is sugar a solid or a liquid?
- What are the properties of a liquid?

Fascinating fact

Solids that are crushed into very small pieces behave like liquids. We can pour sugar but it is still a solid.

1 a Look at the pictures. Is each material a solid or a liquid?

b Follow the lines to discover if your prediction was correct.

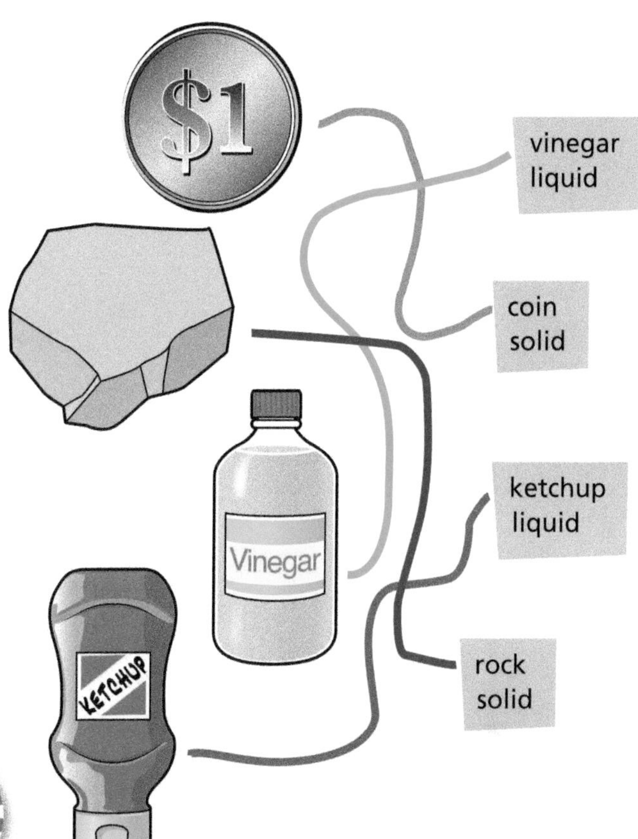

Test yourself

1 Which of these materials are liquids?

a cardboard
b clay pot
c cooking oil
d milk
e saucepan
f water
g wood

12 Water can be solid or liquid

Objective

In this section you will explore how water can be a solid or a liquid.

1 a What are the properties of ice?
b Is ice a solid or a liquid?

REAL WORLD SCIENCE

Scientists study the ice at the South Pole to find out what the climate was like millions of years ago.

Fascinating fact

Water covers 70% of the Earth's surface. That is nearly three-quarters of the planet.

2 a What are the properties of water?
b Is water a solid or a liquid?

3 Make a poster to show the properties of water and ice. Draw a picture of water and a picture of ice. Write the properties around the pictures.

Research task

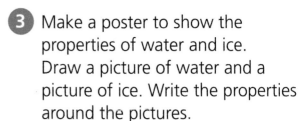

- Find pictures and facts about water and ice to use on your poster.

Test yourself

1 Name a solid.
2 Name a liquid.

13 Changing from solid to liquid and back

Workbook pages 27–28

Objective

In this section you will explore the conditions that make water change from a solid to a liquid and back again.

Talking point

- Have you seen an ice cream melt?
- What else have you seen melting?

Heating and cooling can change the properties of some materials.

1 Look at the pictures below.

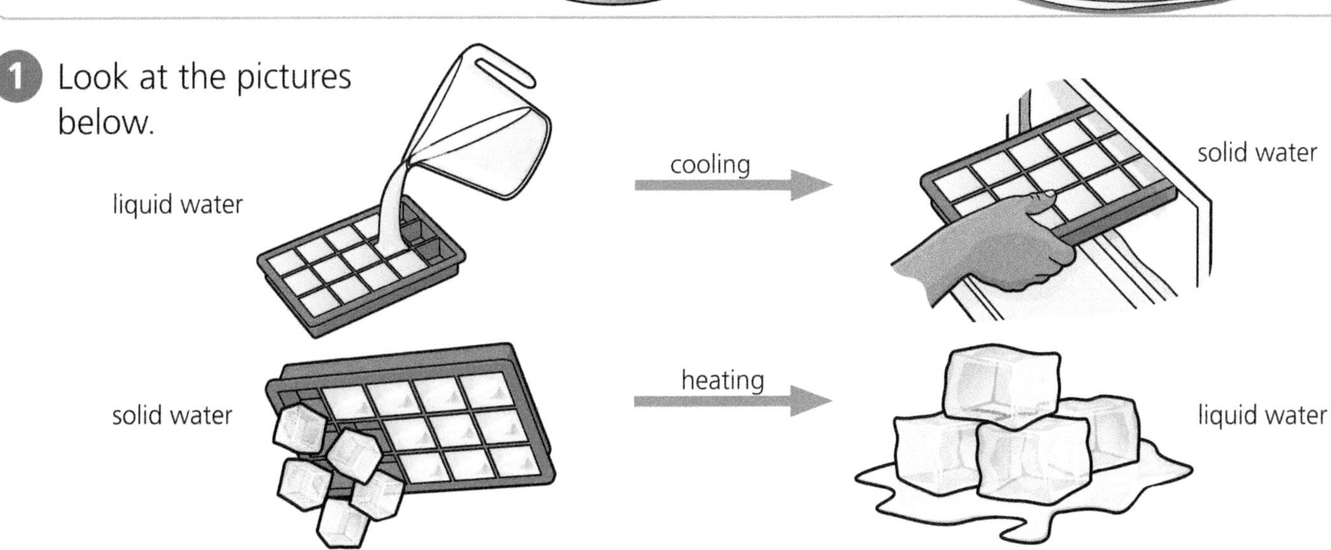

a What happened to the water when it was put in the freezer?

b What happened to the ice cubes when they got warm?

Research task

- Investigate how to keep ice solid for as long as possible.

Test yourself

1 List two properties of a liquid.
2 List two properties of a solid.
3 What happens to water when it is put into a freezer?

14 Liquids are used in the home

Workbook page

Objective

In this section you will find out the names of liquids used at home.

Liquids can be poured, they flow and they fill up the container that they are in.

We use lots of different liquids at home.

1 Look at the pictures of liquids that people use at home.

a Match each liquid to its use. Write your answers in your notebook.

b Write the name of each liquid. Use the words in the word box if you need to.

baby oil cooking oil
fruit juice ketchup
milk vinegar water

Test yourself

1a Write down the names of three liquids you use at home.

b Explain what the liquids are used for.

15 Measuring how liquids flow

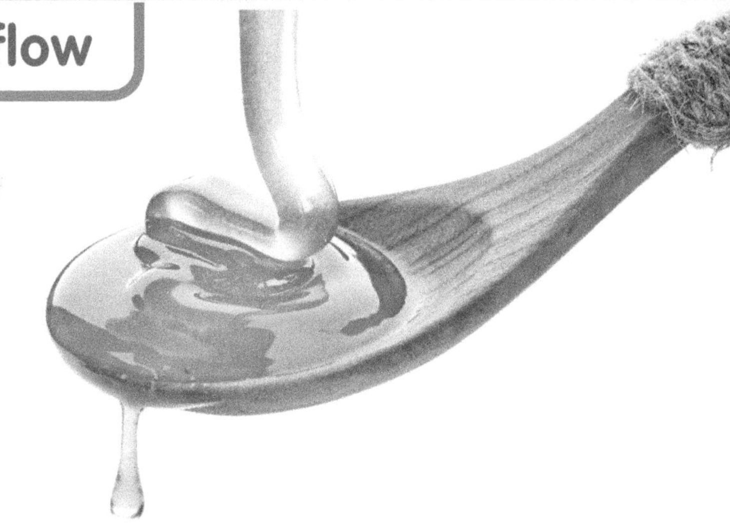

Objective

In this section you will compare the properties of liquids used at home.

Liquids can flow and can be poured. Some liquids flow more easily than others.

Measuring how liquids flow

1 Test four liquids. Work in a group of three.
One person will pour the liquid.
One will time how long it takes to fall off the spoon.
One will record the time.

2 Use a small spoonful of each liquid. Hold the spoon at arm's length above a container. Time how long it takes for the first bit to fall off the spoon.

3 Record your results in a table like this one.

Liquid	Time for the liquid to fall, in seconds
syrup	6

1 Look at the picture.

a Is this material a liquid?

b Does it flow as quickly as water?

REAL WORLD SCIENCE

Builders have to be scientists sometimes. They have to check how runny concrete is before they use it.

Talking point

Talk about your results.

- Do all liquids flow the same?
- Which liquid flowed the fastest?
- Which liquid flowed the slowest?

Test yourself

1 Is this sentence true or false?
All liquids flow as quickly as water.

Investigating what happens to liquids when they are frozen

1 You will test four liquids.
   a Collect some plastic bottles and their tops.
   b Fill each bottle with a different liquid.
   c Label the bottles so you know what is in them.
   d Screw the lids on very tightly so that the liquids do not leak out.
   e Place the bottles in the freezer.
2 What do you predict will happen to the bottles?
3 Look at the bottles the following day.
   What happened to the liquids in the bottles?

STAY SAFE Never touch any liquids without asking first! Do not put any of the liquids in your mouth.

Do all solids melt?

1 Choose three solids to investigate.
   a Choose one that you think will not melt.
   b Choose one that you think will melt.
   c Choose one that you are not sure of.
2 Tell your classmates why you think the solids will melt or not.
3 Place the materials on a paper towel in a warm place.
4 Watch the solids.
   Record what happens to them.
5 Were your predictions right?
6 What happened to the material that you were not sure of?

Unit 3: Check your learning so far

Choose the correct picture in each question.

1 This is how you test if a material is hard.

a 

b 

c 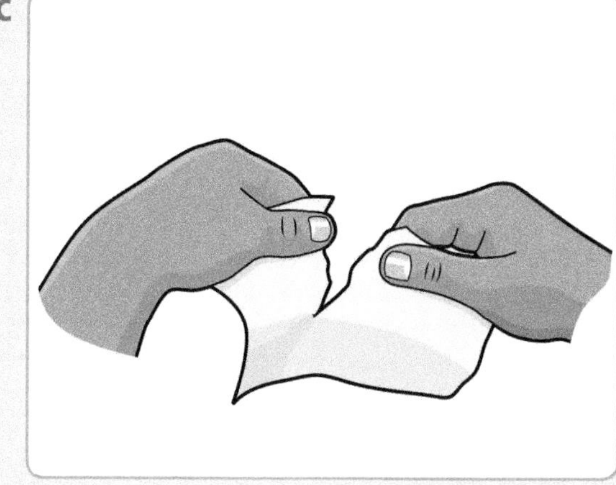

2 This is how you test if a material is strong.

a 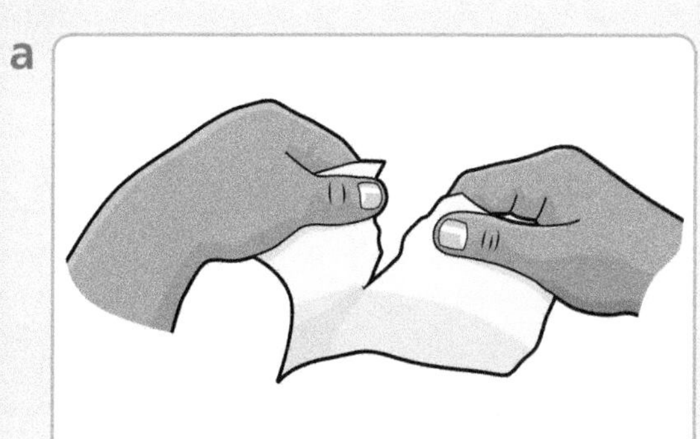

b 

c 

3 Which of these tests will help you find out if a material is a liquid?

a Try to pour the material.

b Try to squash the material.

c Try to scratch the material.

4 Which of these things are liquids and which are solids?

a a coin

b tomato ketchup

c a piece of paper

d vinegar

Unit 4 • The Earth's weather

The weather on Earth is complicated, but we can measure it and learn how to predict it.

KEY WORDS
anemometer
condensation
evaporation
rain gauge
water cycle

Talking point

These two stations are measuring the weather in different parts of the world.

- What differences in the weather do you think they will find?
- What do these symbols mean?

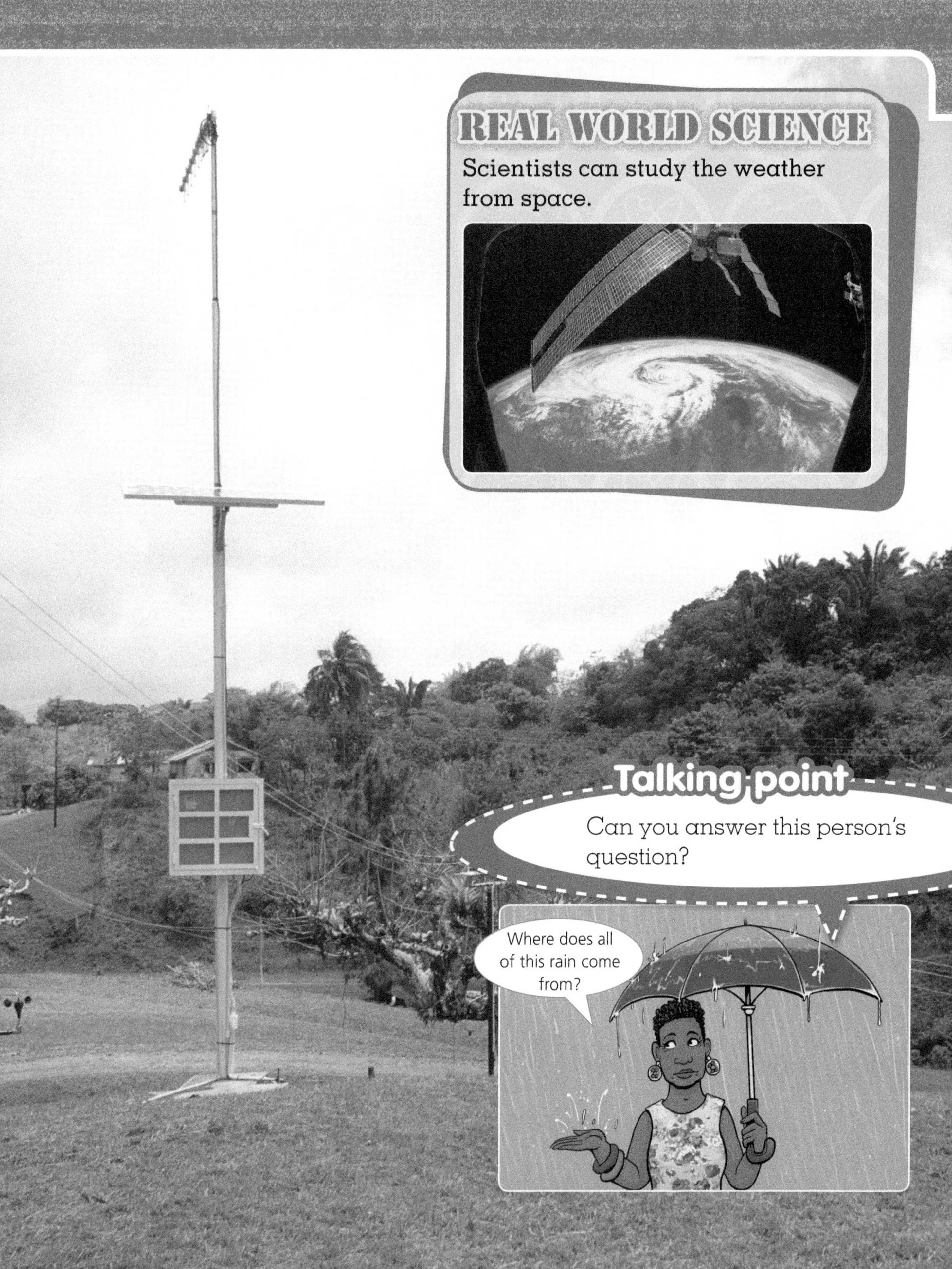

Where does all of this rain come from?

REAL WORLD SCIENCE

Scientists can study the weather from space.

Talking point

Can you answer this person's question?

1 Evaporation and condensation

Workbook page

Objective

In this section you will learn about what happens to water when it evaporates and condenses.

Living things need water. Without water there would be no life on Earth.

Evaporation

Evaporation is what happens when a liquid changes to a gas.

When water evaporates it changes to a gas called water vapour.

Fascinating fact

Most of the rain that falls on your country has come from the sea. The rainwater will run back into the sea or evaporate into the sky.

Measuring and predicting evaporation

You need two identical cups.

1. a Pour the same volume of water into each cup.
   b Measure and record the height of the water.
   c Cover one cup with a lid and leave the other cup uncovered.
   d Leave both cups in the Sun for five days.
2. Predict what you think will happen.
3. Measure the height of the water every day.
4. Was your prediction correct?

Condensation

Condensation is the opposite of evaporation. Water vapour in the air changes from a gas back into a liquid.

Talking point

Compare the photographs on pages 50 and 51. In one picture the water seems to be disappearing.

In the other picture water is appearing.

- Which is which?

Observing condensation

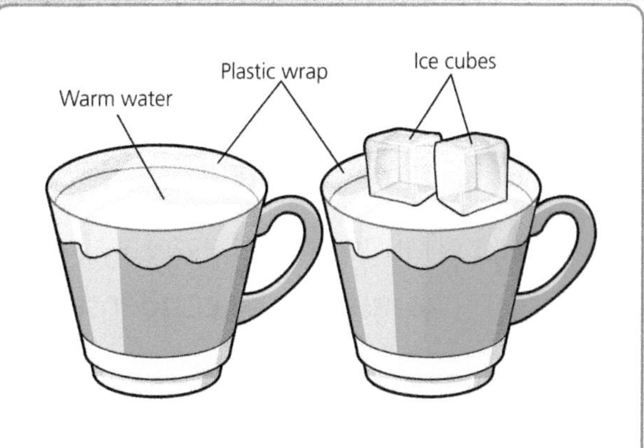

1 a Set up the experiment shown in the picture.
  b Leave the cups on a table in the shade and observe what happens.

2 Did you observe water droplets? Where did you see them?

REAL WORLD SCIENCE

A scientist has invented a device that can take water vapour out of the air to make drinking water. The device uses solar power to cool the water and it can be fixed to a bicycle so that a rider can make their own water as they cycle along.

Test yourself

1 What is the word for water changing to water vapour?

2 What is the word for water vapour changing to water droplets?

2 The water cycle

Workbook pages 31

Objective

In this section you will draw a picture to show the water cycle.

Water evaporates to make water vapour and water vapour condenses to make water droplets.

1 Look at the flow chart below.

a Which change shows evaporation?

b Which change shows condensation?

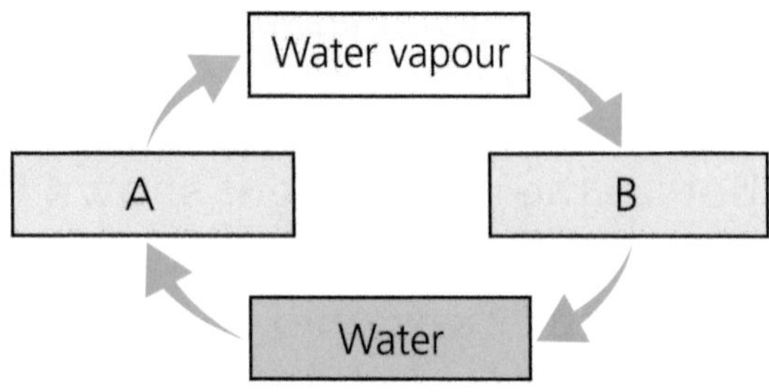

On Earth evaporation and condensation happen on a huge scale. This is called the **water cycle**.

Talking point

- What is helping the water in the sea to evaporate?
- What is helping the water vapour to condense to make rain?

Remember

- Warming up helps evaporation.
- Cooling down helps condensation.

Test yourself

1 Choose the correct word to complete each sentence.

a Water [ ] to make a gas called water vapour.

b Water vapour [ ] to make water droplets.

condenses
evaporates
freezes
boils

3 How much rain is falling?

Workbook pages 33–34

Objective

In this section you will design and make a simple rain gauge.

Look at the pictures of the **rain gauges**. They measure how much rain has fallen.

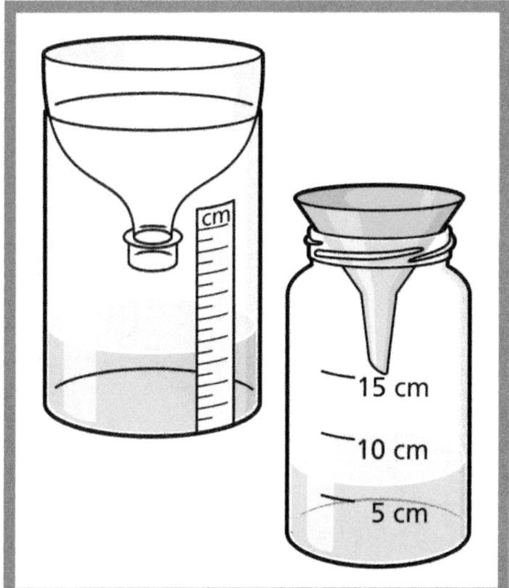

Fascinating fact

A gauge is anything that measures something. Cars have a fuel gauge.

1 a How are the rain gauges catching rain?

b How is the rainfall being measured?

Talking point

- How often should we measure the rainfall?
- Will rainfall be the same all year round?

Measuring the rainfall in your local area

Design and make your own rain gauge.

1 **a** Decide on the best place to put your rain gauge.

b Check your rain gauge every day.

c Record the amount of rain in a table.

2 Which was the rainiest day?

2 What other types of weather could you record?

Research task

- Find out the names of the places with the highest rainfall and the lowest rainfall in the world.

Test yourself

1 Which device measures rainfall?

2 Which units are used to measure rainfall?

4 How windy is it?

Objective

In this section you will design and make a simple anemometer to measure wind speed.

We see evidence of the wind blowing all around us.

Talking point

- Which picture shows the wind blowing very strongly?
- Which picture shows the wind not blowing at all?

The Beaufort scale

The Beaufort scale is a simple way of measuring wind speed using our eyes.

1 Look outside. What number and name do you give the wind today?

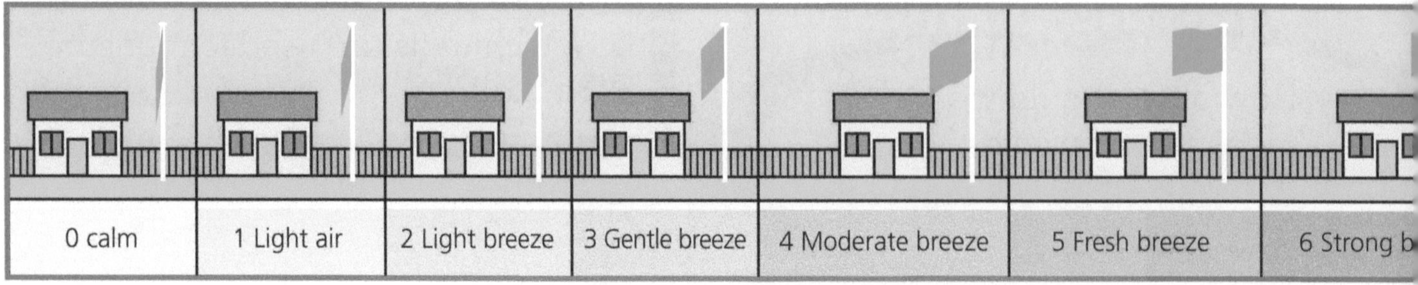

The anemometer

Workbook pages 35–36

Another way of measuring wind speed is a device called an **anemometer**.

Designing an anemometer

You are going to design and make an anemometer.

1 Look at the picture of an anemometer and discuss your plans with a partner.
2 List the materials you will need.
3 Draw your design for your anemometer.

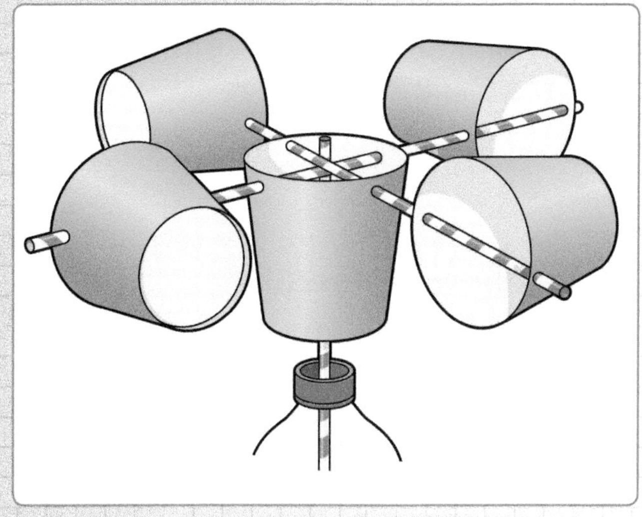

Measuring the wind speed using an anemometer

1 Plan your investigation.
- Where will you place your anemometer?
- How long will you leave it in the wind?
- How will you measure the number of turns of the cups?

2 Use the anemometer to record wind speeds in different places around the school.

The faster the wind blows the more times the cups will turn.

Test yourself

1 What does an anemometer measure?
2 On the Beaufort scale:
  a what is number 6?
  b what is number 10?

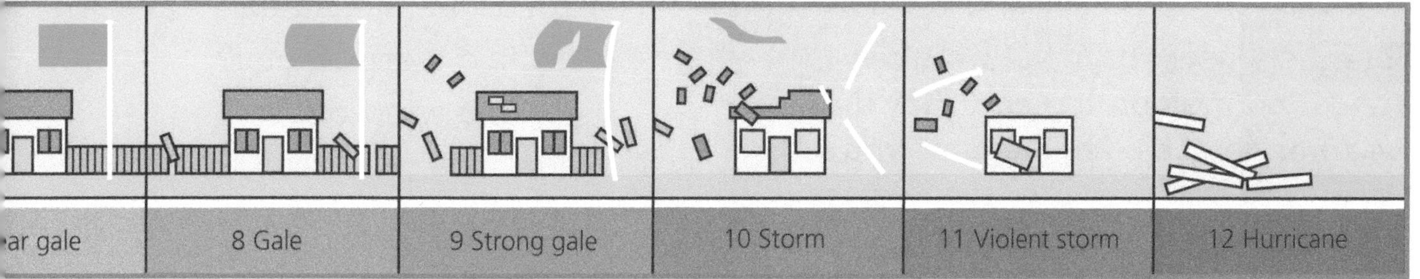

5 Weather charts

Workbook page

Objective

In this section you will learn how to construct a weather chart.

A weather forecaster uses charts and symbols to explain the weather.

Talking point

- Which weather symbols do you recognise?
- What is the weather map showing?

REAL WORLD SCIENCE

Scientists who study the weather are called meteorologists. They use weather devices like the ones you have used and they also use pictures from space.

Fascinating fact

Over 3,000 years ago Babylonians predicted the weather by looking at the shapes of clouds.

1. What can we measure to find out about the weather?
2. a Make your own large weather chart like the one in your workbook. It needs to have spaces to record the weather every day for a week.
   b Display your chart on the wall.
   c Use a rain gauge, anemometer, wind vane and thermometer to measure the weather every day.

Remember

The weather symbols are a quick way to show the weather.
Put these on your weather chart.

Research task

- Find out when your region had its first weather forecast.

Test yourself

1 What do we measure:
  a with a thermometer?
  b with a wind vane?
2 What is the weather symbol for:
  a cloudy?
  b sunny?

6 The weather day by day

Workbook page 38

Objective

In this section you will take weather measurements and compare them.

1 Look at the pupils taking weather readings.

a What are they measuring?

b What devices are they using?

Here are the results they have collected over a week.

	Monday	Tuesday	Wednesday	Thursday	Friday
Wind speed (turns per minute)	10	12	16	35	18
Wind direction	West	West	East	East	East
Rainfall (millimetres)	0	0	12	20	5
Temperature (degrees Celsius)	28	27	30	31	31
Weather symbol					

2 a Which days had the lowest rainfall?

b Which day had the highest rainfall?

3 Was there a link between the rainfall and the wind direction or wind speed?

Talking point

Compare these results with the results from your school.

Test yourself

1 What do these devices measure? Use the word box to help you.

rainfall temperature wind direction wind speed

a thermometer

b anemometer

c wind vane

d rain gauge

Making a permanent weather station

It is important to take weather readings over a long period of time.

1 Design and set up your own weather station at school.
2 Think about how you will protect the instruments but not make them weather proof!

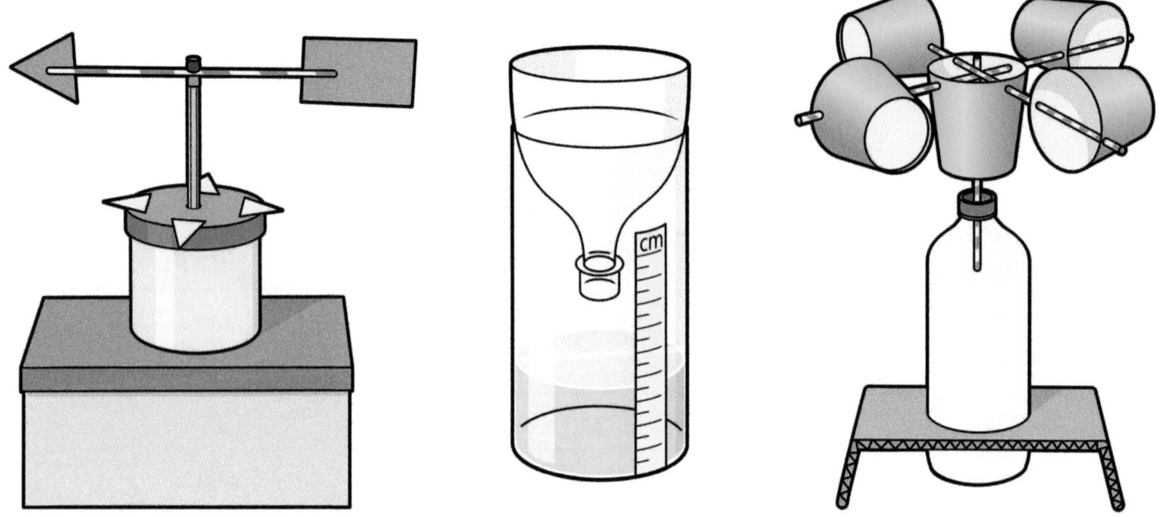

Remember

You will need to take daily readings.

Tracking the weather

1 Check the weather forecast for your area every day for a week. Use the radio, television, newspapers or the internet.
2 The day after the forecast, record what the weather was actually like.
3 Make a display of how accurate the weather forecasts were.
4 Compare your weather readings with the ones you find in the forecasts.

Unit 4: Check your learning so far

Match the description of each device with the correct picture.
Write the name of each device.

1. This device is used to measure temperature.
2. This device is used to measure rainfall.
3. This device is used to measure wind speed.
4. This device is used to measure wind direction.

a
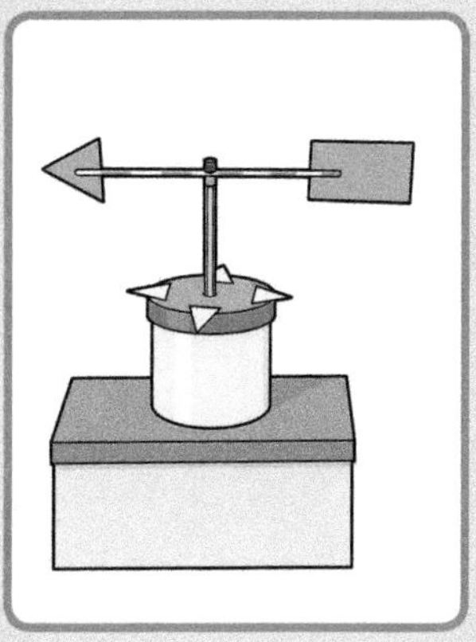

b
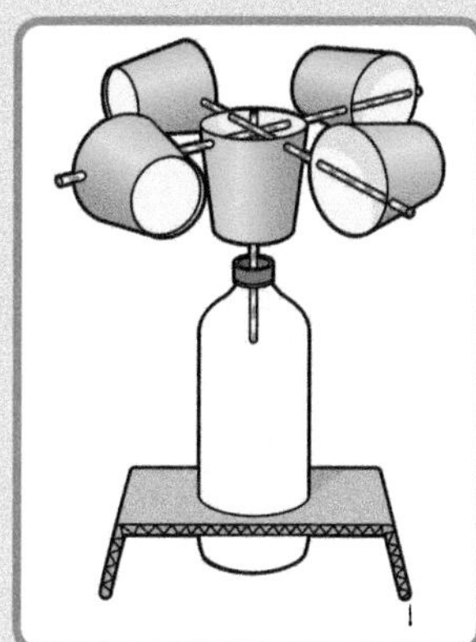

c
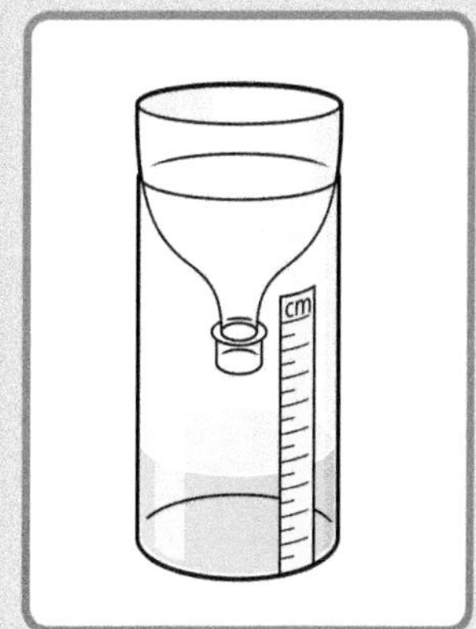

d
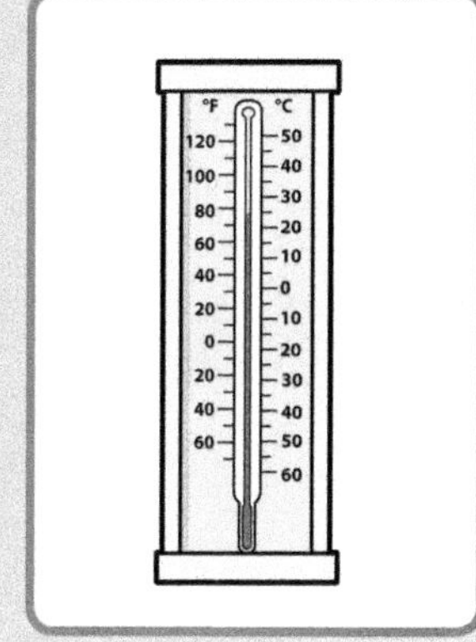

5. Which pictures show evaporation?
6. Which pictures show condensation?

a

b

c

d

7. Draw a picture of the water cycle in your notebook. Label your picture.

Unit 5 • The Solar System

The Solar System is made up of the Sun, Earth and the other planets.

KEY WORDS
Earth
Moon
phases of the Moon
planet
solar energy
Solar System
Sun

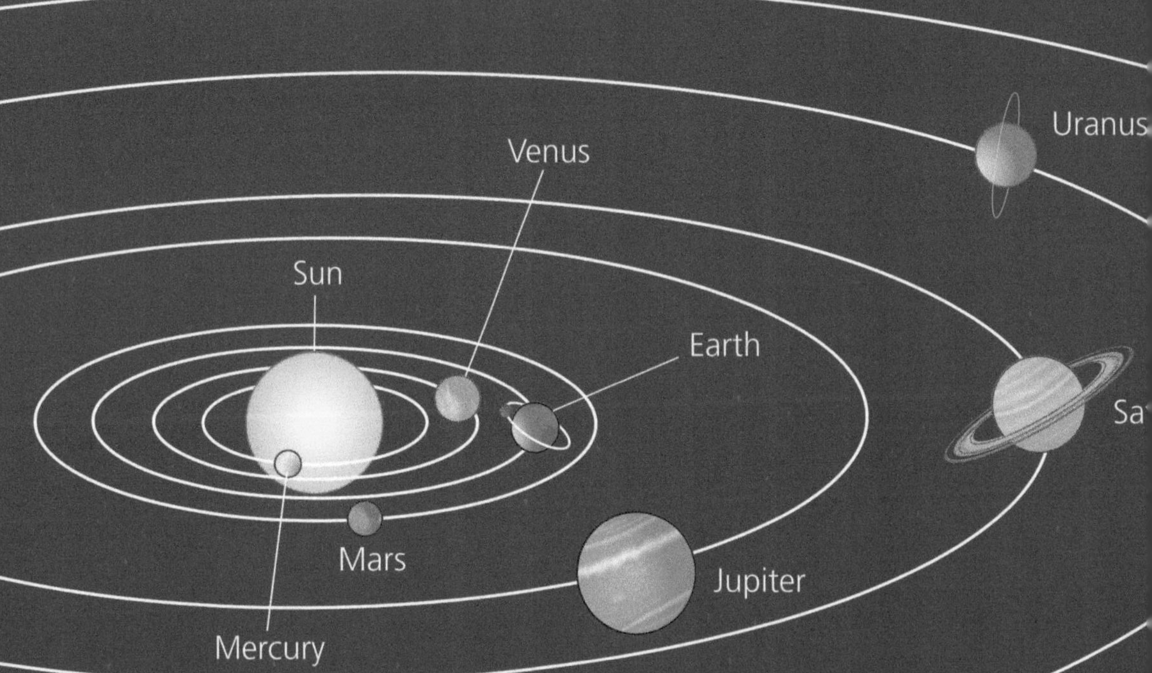

The Sun is at the centre of our Solar System.

Talking point

- Find the Earth in the picture.
- Which planets are closest to Earth?

Look at the picture of the Sun and the planets in our Solar System.

The smallest planet is Mercury.

The largest planet is Jupiter.

Mercury is the closest planet to the Sun. Neptune is the furthest planet from the Sun.

Neptune

Talking point

- Are these pictures of the day or night?
- How do you know?
- How many Moon shapes have you seen?
- Have you ever seen the moon during the day?

REAL WORLD SCIENCE

Astronomers use telescopes to see the stars and other planets.
One telescope is in space. It is called the Hubble Space Telescope.

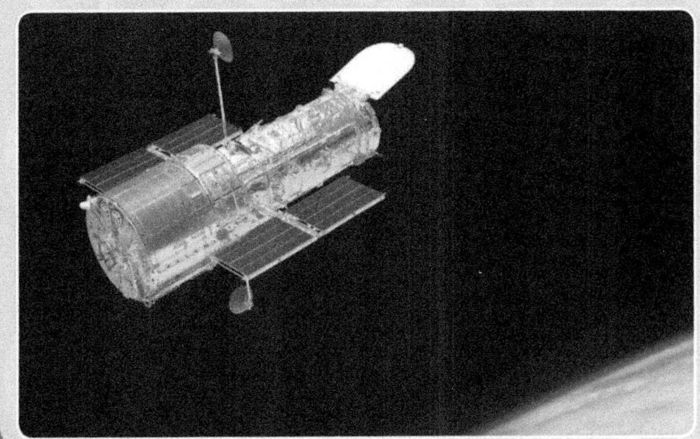

Fascinating fact

Some scientists think they have found a new massive planet on the outer reaches of the Solar System. They are calling it the ninth planet.

The Sun, Earth and Moon

1 The Sun, Earth and Moon are parts of the Solar System

Workbook page

Objective

In this section you will:

- name parts of the Solar System – the Sun, Earth and Moon
- learn that the Sun appears in different positions throughout the day.

Sun

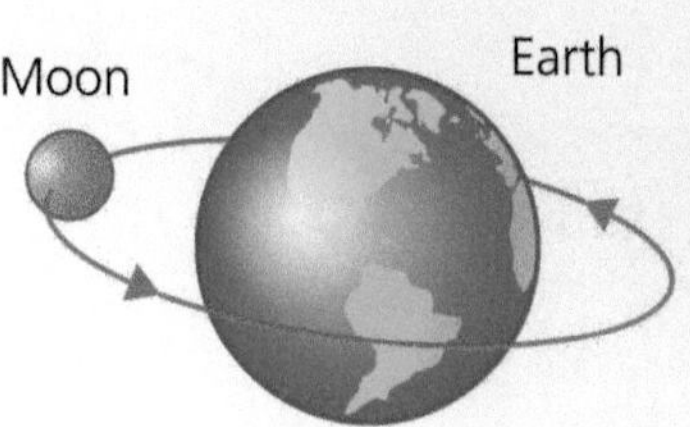

The **Sun** is a burning star. When you go outside you can feel the heat of the Sun on your skin.

The **Earth** is the **planet** that we live on. It is bigger than the **Moon** but much smaller than the Sun.

All three planets are part of the **Solar System**.

The Sun is very far away from Earth.

The Sun is much bigger than the Earth and the Moon.

The Moon is smaller than the Earth and the Sun.

The Moon moves around the Earth and the Earth moves around the Sun.

Fascinating fact

The word solar means 'from the Sun', so **solar energy** is power from the Sun.

The Sun throughout the day

The Earth spins on its **axis** so different parts of the Earth face the Sun at different times of day.

The Earth makes one full turn every 24 hours – this is one whole day and night.

The Sun appears to move across the sky. In fact it is the Earth that is turning.

Talking point

Look at the picture.

- Find the part of the Earth where you live.

1. a When is your part of the Earth in the shade from the Sun?
   b When is your part of the Earth facing the Sun and getting lots of light?
2. Which of these do we call day and which do we call night?

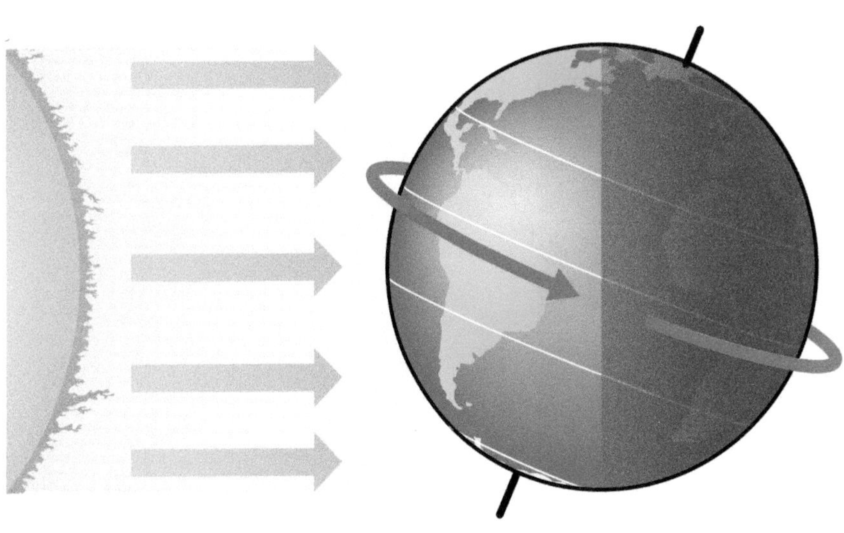

What happens to my shadow in the Sun?

1. Go outside in the morning and look at where the Sun is. Point in the direction of your shadow.
2. Go back to the same place at different times of the day. Point in the direction of your shadow again.
3. Is your shadow in a different direction? Is it the same length?
4. Record your results in your notebook.

Fascinating fact

The Sun is so far away that light from the Sun takes 8 minutes to reach us.

Test yourself

1. Is the Moon bigger or smaller than the Earth?
2. Does the Earth move around the Moon?
3. Does the Earth move around the Sun?

2 The phases of the Moon

Workbook page

Objective

In this section you will label the phases of the Moon.

Fascinating fact

The Moon is like a mirror. It reflects the light from the Sun.

As the Moon moves around the Earth, part of the Moon is in sunlight and part of the Moon is in shade.

The Moon seems to change shape because we only see the part that is in sunlight.

The different shapes we see are called the **phases of the Moon**.

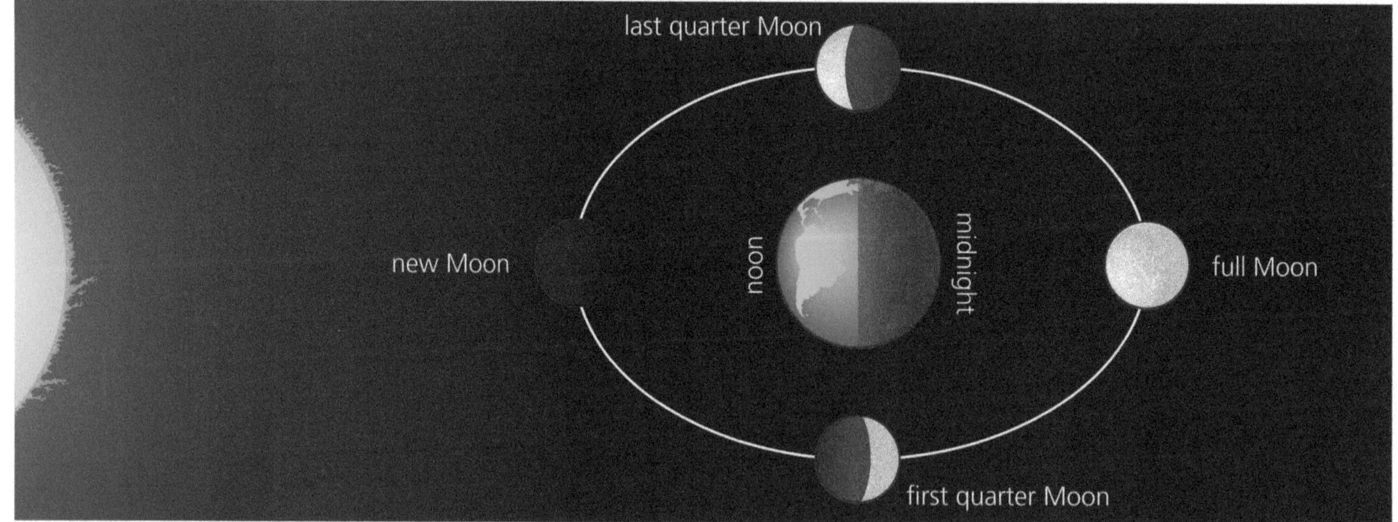

Phases of the Moon game

Play the phases of the Moon game with a partner.

1 Cover the picture above.
2 One person says the name of one of the phases and the other person tries to draw it.
3 Now look at the picture. Did you get it right?
4 Now let the other person try.
5 Continue until you have both drawn three phases of the Moon.

Test yourself

1 Name three different phases of the Moon.
2 Explain how we see a full Moon.

3 Day and night

Objective

In this section you will identify patterns in day and night.

Sometimes the part of the Earth where we live is facing the Sun. Sometimes it is facing away from the Sun, so it is in the shade.

The Sun rises in the east and sets in the west.

Sunrise is the start of the day.
Sunset is the end of the day and start of the night.

Recording day and night where you live

Listen to the weather report every day for five days.

1 Find out the times of the sunrise and sunset where you live.
2 Record the times in a table.
3 Does the time of the sunrise change?
4 Does the time of the sunset change?

Talking point

Point in the direction of the west.

- Look there this evening.
- What will you see?

1 Look at the table. It shows the sunrise and sunset times for London.

Time of year	Sunrise and sunset times for London, England	
	Sunrise	Sunset
1st December	07:43	15:54
1st June	04:50	21:06

a How long is the daytime at the beginning of December?

b How long is the daytime at the beginning of June?

Talking point

- How long is the daytime where you live?
- Does it change over the year?

Test yourself

Copy and complete the sentences.

1 Sunrise is the ☐ of the day.
2 Sunset is the ☐ of the day.

Sunrise and sunset

1 Research the sunrise and sunset times of another country. You can do this by listening to the weather reports for the country or using the internet.
2 Record the time of sunrise and sunset over a month.
3 Compare these times with another country.
4 Design a table to record and display your results.

Making a model of the Sun, Earth and Moon

1 Use balloons of different sizes to make papier mâché models of the Sun, the Earth and the Moon.
2 Look at the picture on page 60 to check the sizes of the different planets and the Sun.
3 Blow the balloons up and cover them in strips of paper dipped in a flour and water paste.
4 When the papier mâché is dry, pop the balloons.
5 Paint your models so they look like the Sun, the Earth and the Moon.
6 Try to find out the distances between the Sun and the Earth.

Unit 5: Check your learning so far

Choose the correct picture to match each description.

1 This is the Moon.

2 This is the Sun.

3 This is the Earth.

a 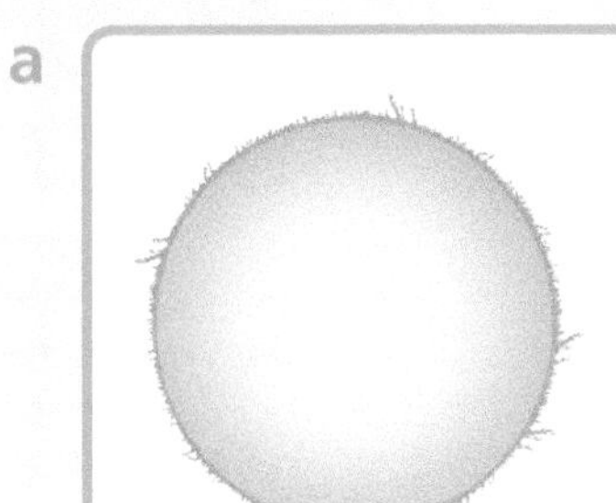

b 

c 

4 This is the position of the Sun at midday.

5 This is the position of the Sun in the morning.

6 This is the position of the Sun in the evening.

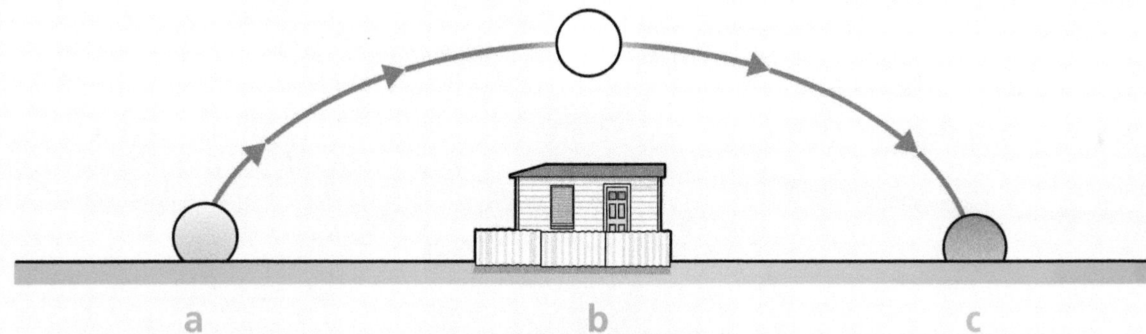

a b c

7 This phase of the Moon is a quarter.

a 

b 

c 

Unit 6 • Ecosystems 2

All life on Earth is linked together. Living things depend on each other and on their environment.

KEY WORDS
carnivore
environment
herbivore
omnivore
predator
prey

Adaptations help animals and plants to live and survive in their habitat.

Tyrannosaurus rex was a fierce dinosaur. This one is hunting.

Talking point

- What does the T. rex want to eat?
- How is the T. rex adapted to do this?

Fascinating fact

The last dinosaurs lived on the Earth over 65 million years ago. Many scientists think they died out after an asteroid or meteorite collided with the Earth near the Caribbean. We will not meet any dinosaurs, but we can meet their relatives – the snakes, lizards and birds.

Talking point

- What is the triceratops eating?
- How can the triceratops protect itself?
- How is the triceratops adapted to eat plants?

Talking point
Animals that live on the Earth now eat plants or other animals. Some eat both.
Which of these animals:
• eat other animals?
• eat plants?
Talking point
Look at the picture of the woodland.
• How are humans damaging this environment?
• What will happen to the animals and plants?
• What is causing a lot of noise?
• Why might noise be a problem?

Feeding relationships and defence in animals

1 Eating and being eaten

Objective

In this section you will:

- find out what producers and consumers are
- learn about predators, their prey and what links them
- learn how animals are adapted to eat and to avoid being eaten.

Talking point

Look at the picture of animals and plants that live in a pond.

- Which animals might eat the plants
- Which animals might eat other animals?

We can show what eats what in a food chain. A chain always starts with a plant.

Green plants make their own food. They use energy from the Sun. Plants are called **producers**.

Animals cannot make their own food. They have to eat plants or other animals. Animals are called **consumers**.

Consumers can be primary, secondary or tertiary consumers.

An animal that is eaten by another animal is the **prey**. The animal that eats it is the **predator**.

Workbook pages 41–42

Food chains

In the food chain:

- is the caterpillar the predator or the prey?
- is the bird the predator or the prey?

1 Think of an animal that might eat the bird.

The animal that eats the bird is a tertiary consumer. That animal is the third consumer in the food chain.

Animals that eat plants are called **herbivores**.

Animals that eat other animals are called **carnivores**.

Animals that eat plants and animals are called **omnivores**.

2 Is the caterpillar a herbivore, an omnivore or a carnivore?

3 A human can be a primary, a secondary and a tertiary consumer in a single meal. Explain how this is possible.

Defence against being eaten

Animals and plants protect themselves from being eaten.

4 a Look at the pictures. Which method of protection does each animal use?

spines	sting	venom

b How do the protection methods work?

5 Why does this plant have spikes?

Test yourself

1 Why are plants called producers?

2 Name an example of a primary consumer.

2 What is environmental destruction?

Workbook page

Objective

In this section you will:
- find out what 'environmental destruction' means
- investigate what causes environmental destruction.

Fascinating fact

The word 'bio' means 'life'.

The **environment** is everything that surrounds living things.

In an environment the living things are called the biotic parts. The non-living things are called the abiotic parts.

Talking point

- How do these things cause destruction of the environment?
- Which have you seen in your local area?

1 Look at the picture of the coral reef.

  a What clues are there that this environment is healthy?

  b Find some biotic parts.

  c Find some abiotic parts.

Test yourself

1a List two abiotic parts of an environm

  b What are the biotic parts?

2a How do humans destroy environme

  b What happens to the plants and animals when this happens?

3 Looking after the environment

Workbook pages 44–45

Objective

In this section you will identify how we can stop environmental destruction.

Remember

The environment is everything that surrounds living things.

Everything in our environment is linked.
Something that happens in one place can have a big impact in lots of other places too.

Local people are angry that a local beauty spot has been destroyed. Soil and **waste** has washed onto a beach well known for its soft sand and clear water.

'This is terrible!' said Mrs. Greenidge. 'This mess will keep tourists away and cost me money.' Another local person, Mr. Ward, agreed: 'My children used to swim and play here but now it is unhealthy.'

A spokesperson for a local wildlife society was also worried. 'This damage will kill many fish and birds. The water is too cloudy and their food is gone,' said Dr. Alleyne. 'The soil has come from hills where trees have been cut down. Where will the animals that lived in the trees go? Soil erosion and throwing waste away are big problems.'

Mr. Browne has been blamed for destroying the beach environment. He said it wasn't his fault: 'We needed the trees for building and for firewood. It is my land and the steep hillside was the best place to cut the trees. There is nowhere else I can throw my garbage either. I don't see how this can damage a beach hundreds of metres away.'

1 Listen to the newspaper report being read out. Follow the words on page 73–74.

a What is causing the environmental damage?

b What impact is the environmental damage having on wildlife?

c What impact is the environmental damage having on people?

You are part of a team of scientists trying to solve the environmental problem.

1 Explain to Mr. Browne that cutting down the trees and throwing away the waste can destroy an environment.

Make sure you:

- explain how trees can stop soil washing away
- explain how waste can roll down hills and be washed away by rain.

2 Present your ideas to the class.

Research task

Pollution from cruise ships is a major source of environmental damage.

- Find out what the pollution from the cruise ship is.
- What could be done to reduce this pollution?
- Write a letter to the person who owns the cruise ships telling them about the problems.

4 How should we treat solid waste?

Objective

In this section you will:
- learn more about solid waste
- learn how we manage solid waste.

We all create waste at home. We use things and there is waste left over.

Talking point

Look at the picture.
- Which of these objects do you use at home?
- How do you get rid of the waste when you have used these objects?

Workbook page

Handling solid waste

Waste needs to be handled properly or it will damage the environment. Handling garbage properly is called **solid waste management**.

Research task

- Find out what happens to the garbage in your local area after it is collected.
- What kinds of garbage are recycled, reused, buried or burnt?

There is probably a solid waste management department in your community.

We all have to play a part in managing solid waste.

Talking point

- How can you produce less waste?
- What things do you use and how do you use them?
- How much packaging is there on the things that you use?

1 What did you do with the last plastic bottle you used?

Fascinating fact

Ten litres of water is needed to make just one sheet of A4 paper!

Test yourself

1 Write down three examples of solid waste.

2 Describe two roles of the solid waste management department in your community.

5 Cleaning up litter

Objective

In this section you will learn more about litter in schools and clean-up projects.

Litter is solid waste that has been thrown to the ground and left there.

Litter damages the environment in lots of ways.

Litter attracts insects, rats and birds that can spread disease.
Objects in litter can cut and trap animals.
Making paper and packaging causes pollution.
Litter rots and produces gases that can add to global warming.
STOP

What can we do to help manage waste in our school?

My clean-up project

1 Safely collect litter that has been dropped in the school grounds for ten minutes.

2 Make a list of the different objects you find.

3 Place the litter onto weighing scales or a top-pan balance.

 a How many grams did you collect?

 b How many grams did your class collect?

STAY SAFE Check with your teacher before picking anything up.
Wear gloves and wash your hands after your survey.

Test yourself

1 Describe two problems caused by litter.

2 Suggest two ways of reducing litter.

6 The sound of music

Workbook page

Objective

In this section you will learn how your ears hear different sounds.

Talking point

- What sounds do the things in the picture make?
- How are the sounds different?
- Are they loud or quiet? Low or high?

The ear

The ear is the **organ** that detects sound. The picture shows how sounds travel to our ears. The tiny eardrum in each ear **vibrates** and we hear the sound. All sounds are made by vibrations.

1 Why do some animals have large ears?

Sounds can be soft or loud. This is the volume.

2 Name an animal that makes loud noises.

Sounds can have a low pitch or a high pitch. A large instrument such as a bassoon makes low-pitched sounds. A small instrument like a piccolo makes high-pitched sounds.

Investigating pitch

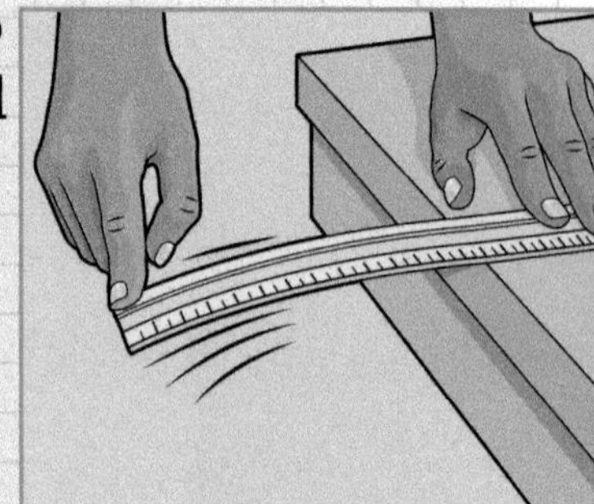

1 Place a ruler so it is half on and half off your desk.
2 Make the ruler vibrate.
3 What happens to the sound when you move the ruler further onto the desk or further off the desk?

Test yourself

1 Which part of our body allows us to hear sounds?
2 Which animal makes a noise with a lower pitch, a cow or a mouse?

7 Making a musical instrument

Workbook pages 48–51

Objective

In this section you will make and use simple musical instruments.

Different musical instruments make different sounds.

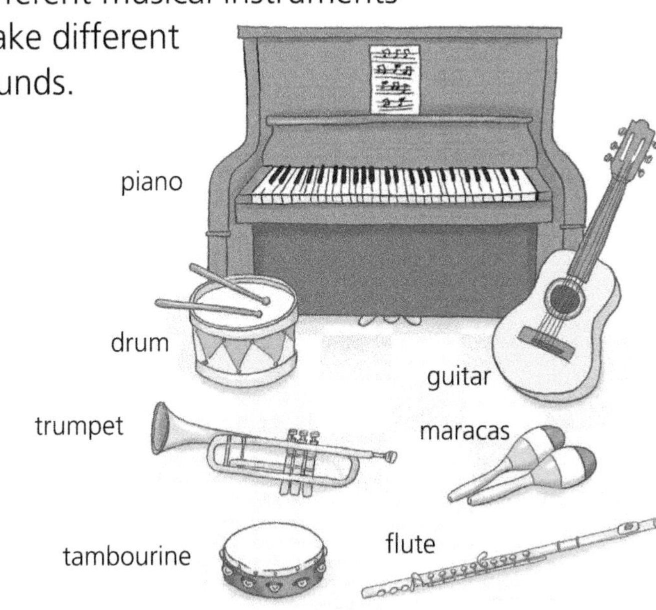

Have you heard any of these instruments played? They all make different kinds of sound.

Talking point

Which instrument makes:

- the loudest sound?
- the sound with the lowest pitch?

1. What is the difference between a noise and music?
2. Do all people like the same sounds?

Making a musical instrument

1. Choose some materials and make your musical instrument.
2. Use your imagination.

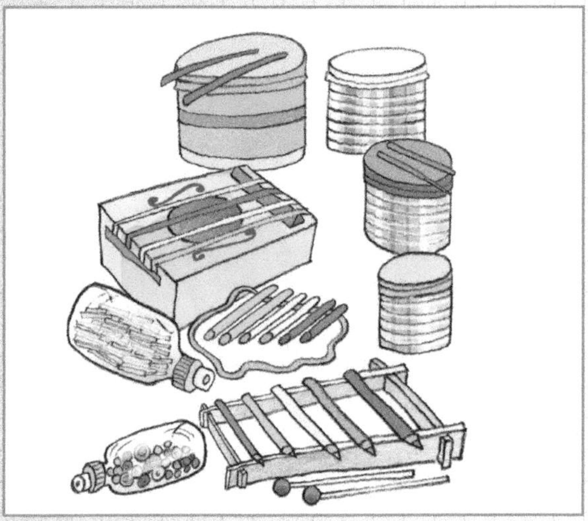

3. What different sounds can your instrument make?
4. Can your class make an orchestra and play a tune?

Test yourself

1. The loudness of a sound is called its volume/intensity.
2. Whether a sound is high or low depends on its distance/pitch.
3. All sounds are made by vibrations/volume.

Making a model of soil erosion

1. Make this model to investigate what happens when water runs down a slope.
2. Pour water from a jug or watering can down the slope. This is modelling rain. Try light rain and heavy rain.
3. a Now add some twigs. These are modelling trees.

   b Do the 'trees' make any difference?
4. Try some other ways to slow down soil erosion.

Investigating sound

1. Make a guitar. Stretch rubber bands of different thicknesses around a piece of cardboard.
2. Does the thickness of the rubber band make any difference to the sound?
3. Change the length of the bands by holding them down in the middle.
4. Does changing the length change the sound? Explain how the sound changes.

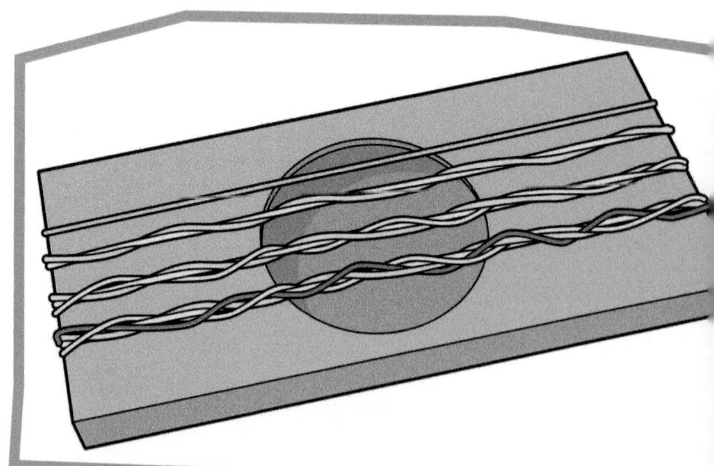

Remember

Think back to your investigation with the vibrating ruler.

Unit 6: Check your learning so far

Choose the correct picture to answer each question.

1 Which living thing is a producer?

2 Which living thing is a primary consumer?

a  b  c  d 

3 Which living thing is a predator?

4 Which living thing is a herbivore?

a  b  c  d 

5 Read the sentences. What is each sentence describing? Choose the correct answer from the word box.

a This can cause environmental destruction.

cleaning up litter	cutting down a forest

b These are a big cause of pollution.

cruise ships	trees

c We use this to hear sounds.

ear	mouth	nose

Unit 7 • Forces, motion and structures

Pushes and pulls can make objects move, stop, change direction or change shape.

KEY WORDS
device
mechanical
motion
pull
push
start
stop

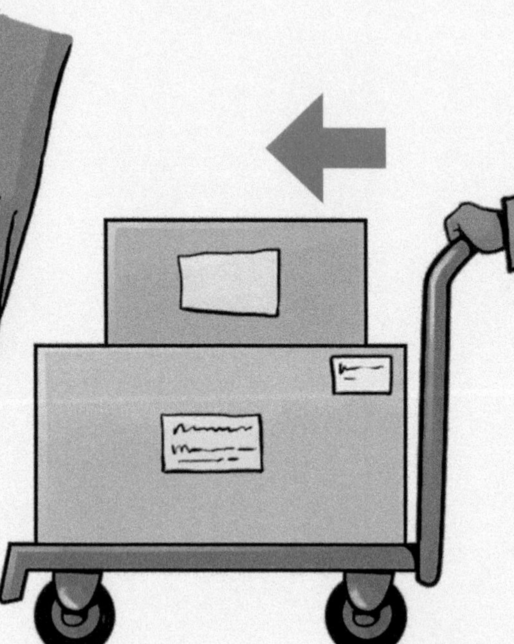

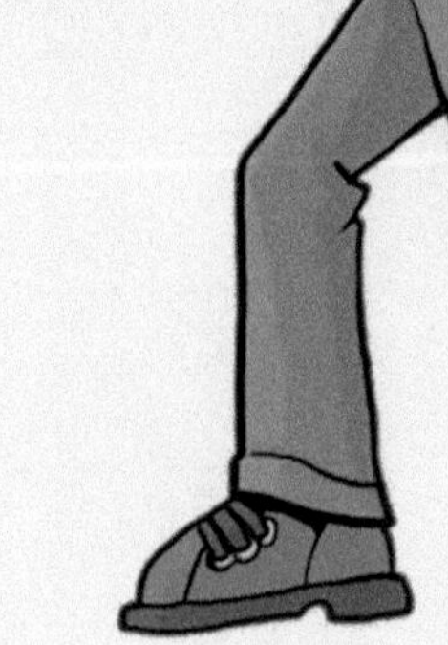

Talking point

Look at the pictures. What is:

- being pulled?
- being pushed?
- changing direction?
- being stopped?
- changing shape?

Talking point

- What are these small devices used for?
- How do they help us to do jobs?
- How do some of these devices hold a bicycle together?

Fascinating fact

A modern racing bicycle weighs only 7 kilograms. All of the parts are made of light material such as carbon fibre.

REAL WORLD SCIENCE

Scientists use their skills to design parts for racing cars. Special brakes slow the cars down from 300 kph (kilometres per hour) to a stop. The front of a racing car is designed to crumple up in a crash. This protects the driver.

Effects of forces

1 Using forces in sports

Workbook page

Objective

In this section you will understand that forces can make things move.

Pushes and **pulls** can make objects move.
This movement is called **motion**.

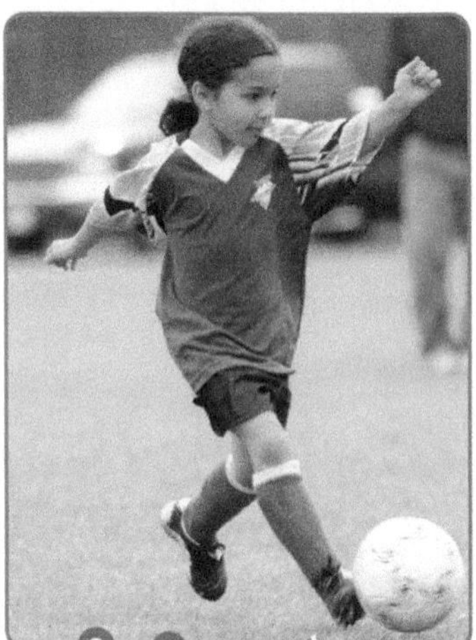

Talking point

- What is this child doing?
- Is the child using a push or a pull force?
- What is happening to the ball?

Using forces to make objects move

1 Put a ball on a flat surface outside. Make sure it is completely still.
2 You can make the ball move. Talk about the different ways you could do this.
3 Test some of the different ways of making the ball move.
4 Identify if you are using a push or pull force each time.

Fascinating fact

When a top golfer hits a golf ball it can speed up from 0 kilometres per hour to over 350 kilometres per hour!

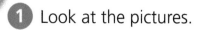

1 Look at the pictures.
   a Is the person using a push or a pull force to hit the baseball?
   b Is the person using a push or a pull force to throw the basketball?

2 Try these for yourself if you are not sure

Test yourself

1 How did you **start** a ball moving?
2 Did you use a push or a pull force?

2 Changing direction

Objective

In this section you will explore how moving objects can change speed, direction or stop.

Talking point

- How can you **stop** a rolling ball?
- Discuss the different ways you could do this.

Predicting what happens in a collision

1 Predict what will happen to the cars when they hit each other.
   a Will the cars stop?
   b Will they both travel in the same direction?
2 Use two toy cars of the same size to test your prediction.

1 Test what happens if one of the cars is much bigger and heavier than the other one.

Changing the direction of a moving object

1 Roll a toy car across a table or the floor.
2 Make the car move to the right and then to the left.
3 What force are you using?

Directing a toy car using only pushes and pulls

Sit facing your partner.

1 Push the car towards your partner.
2 Your partner has to use pushes or pulls to make the car come back to you.
3 You are not allowed to stop the car or pick the car up.

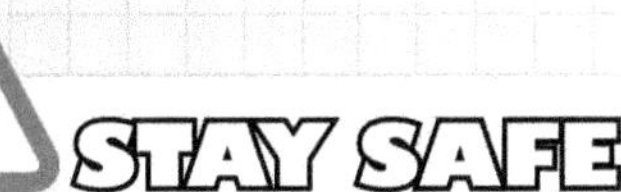

Do not push the car too hard or too fast.

3 Mechanical devices around us

Workbook pages 5

Objective

In this section you will:

- identify some simple mechanical devices
- learn what some simple mechanical devices are used for.

We use simple **mechanical devices** to hold things together.

The smallest devices hold the biggest structures together. Look at the bolts on this bridge.

Talking point

Think of three mechanical devices that you have learnt about before.

Simple mechanical devices can do lots of jobs for us.

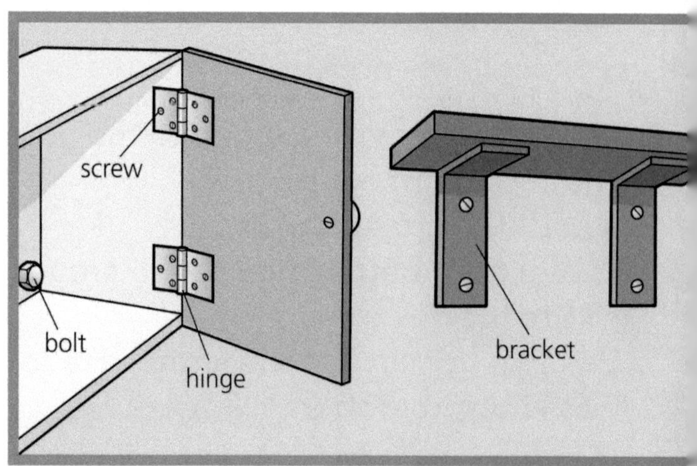

1. Look at the picture of the cupboard.
   a What are the screws used for?
   b What would happen if the screws were removed?

The hinges fix the door to the cupboard and allow the door to open and close. Brackets fix a shelf to the wall.

2. Can you see any brackets being used in your classroom?

Mechanical devices in the classroom

1 Now look for small mechanical devices in your classroom.
   a How many nails, bolts and screws can you find?
   b Record where you find them and what they are being used for.

2 Look around your classroom again.
   a Find some hinges.
   b Record where you find them and what they are used for.
   c What would happen if these hinges were removed?

3 In your notebook, write the name of each simple mechanical device.

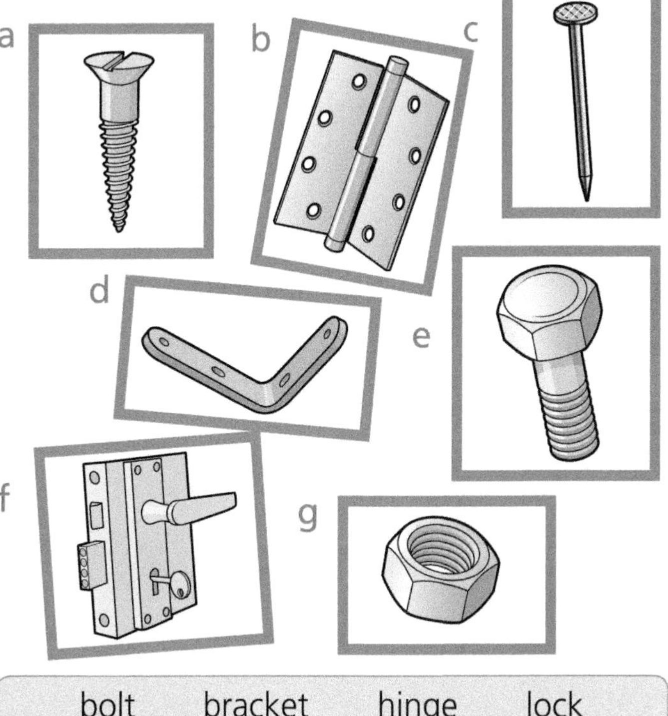

bolt	bracket	hinge	lock
nail	nut	screw	

Fascinating fact

Nails are a type of wedge. Metal nails have been used for over 3000 years!

Devices have special shapes to help them do their job.

Why are nails the shape they are?

Try to push a pencil into modelling clay.

1 Which way round is it easiest to push?
2 What does this tell you about the shape of nails?

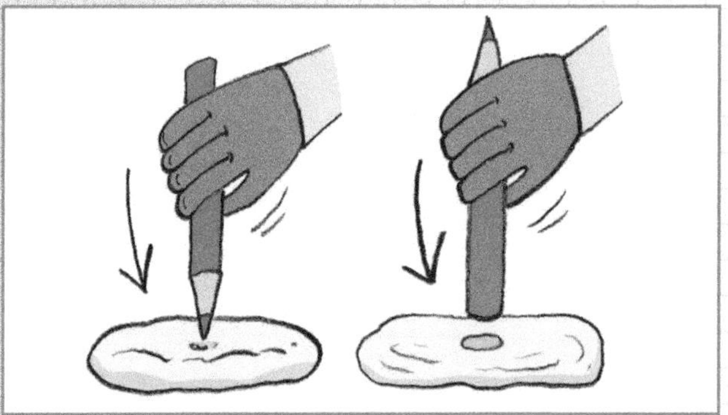

Test yourself

1 Name three small devices used for holding objects together.
2 Describe an example of a bracket being used.
3 Explain why we put hinges on doors.

Making a model

1 Use a modelling set, such as Meccano®, to investigate how screws, nuts and bolts are used.
   a Practise using the screwdriver or spanner to attach the screws or bolts to the parts of the structure you are making.
   b Practise pushing the bolts into the structure and securing them with the nuts.
2 Try to build a structure using these simple mechanical devices.

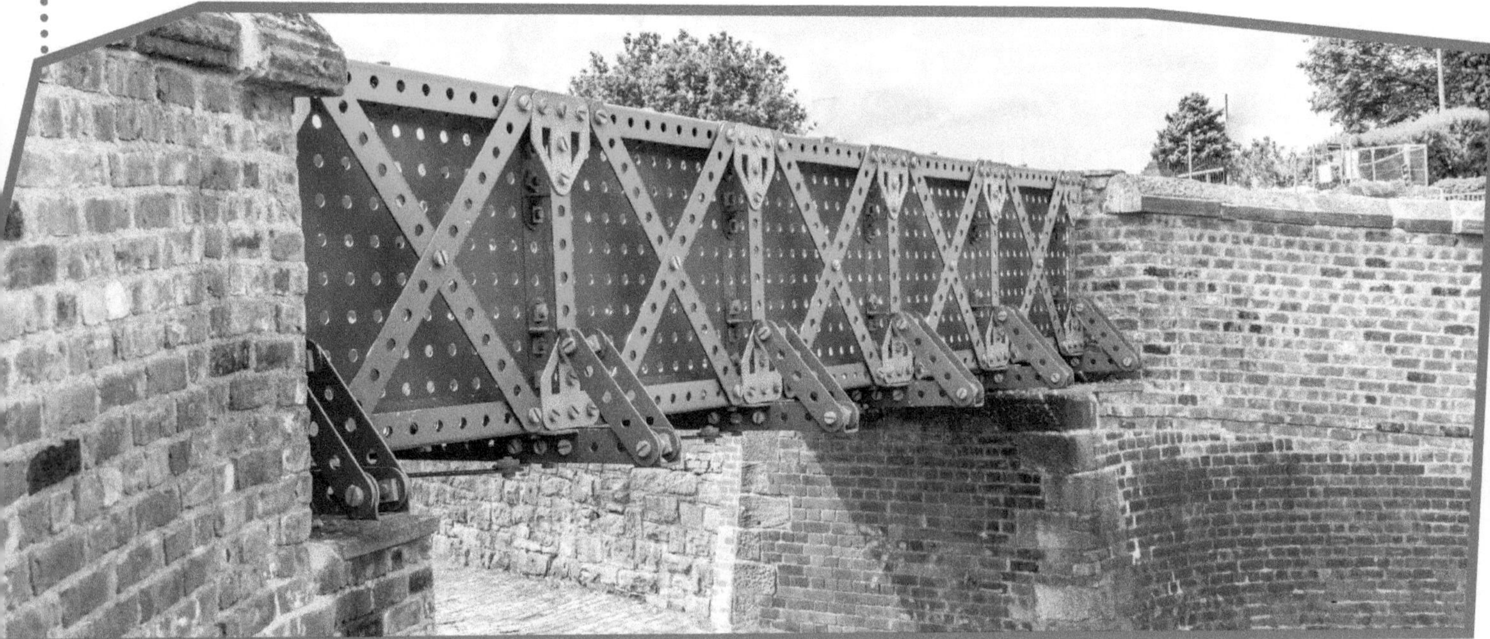

Building a zoo or village using some mechanical devices

Decide what you want to build in your class. You could construct a village or a zoo, for example.

1 Decide on the structure you are going to build.
2 How will you join the parts of the structure together?
3 a Choose the materials from the ones you are given.
   b Use cardboard, paper, paper fasteners, paper clips, sticky tape and straws.
4 Make a display with the other pupils in your class of all the structures you have made.

Unit 7: Check your learning so far

1 Which picture shows a person using a push force?

a 

b 

c 

2 Which picture shows a person using a pull force?

a 

b 

c 

3 If you push a car on its right-hand side, it will move …

a forwards.
b backwards.
c to the right.
d to the left.

4 Look at the pictures of simple mechanical devices.

a Which picture shows a nail?
b Which picture shows a screw?
c Which picture shows a bracket?

a 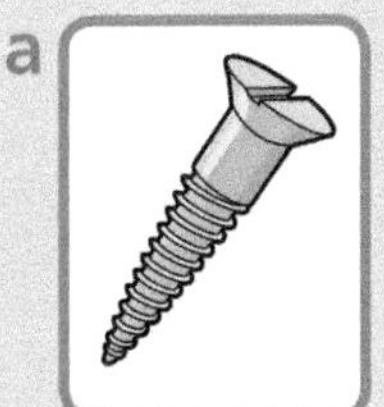

b 

c 

d 

e 

f 

g 

Unit 8 • Diversity and classification

All living things are different but they have some important things in common.

KEY WORDS
alike
different
feeding
growth
living
non-living
reproduce

Talking point

These animals and plants are all different. Some do look similar though.

- Group the animals and plants so the similar ones are together.
- Which things in the pictures are non-living?
- How can we tell living things from non-living things?

REAL WORLD SCIENCE

Some scientists studied 17 trees in the rainforest. They found 1,200 new species of beetle never before known about. New species are discovered every day. Sadly, other species become extinct every day.

Fascinating fact

Dolphins and whales look like fish but they should be grouped with humans. This is because they are mammals.

Talking point

These children are all the same age.

- How are the children similar?
- How are the children different?

Human variation

1 The same and different

Workbook pages 5

Objective

In this section you will describe the ways that people are alike and different.

We use differences to help us to recognise other people.

People vary in height and body shape, skin colour, eye colour, hair colour and whether they are male or female. People also have different-shaped faces.

Talking point

- How are these people the same?
- How are these people different?

1. a Who is this person?
   b How did you recognise him?
2. a On a piece of paper, draw the face of a person you know well.
   b Colour your picture.
   c Show your drawing to a partner. Can they work out who it is?

Test yourself

1 List two features you could use to describe a person.

2 How are we the same and different?

Workbook pages 57–58

Objective

In this section you will group yourselves according to similarities.

Your height when you are an adult will not be the same as it is now. All people have **growth** spurts and the shortest person in a class might become the tallest adult.

Research task

- Find out who is the tallest person in the world.

A class comparison

1 Draw a picture of yourself.
- If you are tall, make your drawing fill the whole page.
- If you are not very tall, only use some of the page.

a Colour your skin and hair.
b Add other important features so the picture really looks like you.
c Ask someone to measure your height. Write this at the bottom of your picture.

2 Display your pictures in order of height. Place the tallest person at one end and the smallest person at the other end.

a Place your picture where you think it should be.
b Discuss this with your classmates around you. Change the order if you need to, then stick the pictures in place.

Test yourself

1 Name a piece of equipment you can use to measure height.
2 Describe two things you have in common with the rest of the pupils in your class.
3 Describe two differences between you and another pupil in the class.

3 Alive or not alive?

Objective

In this section you will be able to describe examples of living and non-living things.

Living things can breathe, move, eat, grow and **reproduce**.

1 Look at the picture of the beach and sea.

a Find some examples of living things.

b Find some examples of **non-living** things.

c What clues helped you decide if the things are living or non-living?

Fascinating fact

Some things are classed as 'once-living'. The wood used in furniture was once part of a living tree but it is dead now.

2 List three things that are once-living.

3 Look at the list of words. Which of these abilities show that something is living?

breathing changing colour falling feeding growing heating up moving reproducing turning

Research task

- Coral doesn't move very much but it is alive. Find out why a coral is a living thing.

Test yourself

1 List three things that living things can do that non-living things cannot do.

2 Explain what a once-living thing is.

3 Explain why trees are living even though they do not move around very much.

4 Living and non-living display

Workbook pages 59–60

Objective

In this section you will make a presentation showing living and non-living things.

This class has made a display of spinning plates. They are presenting what they know about living and non-living things.

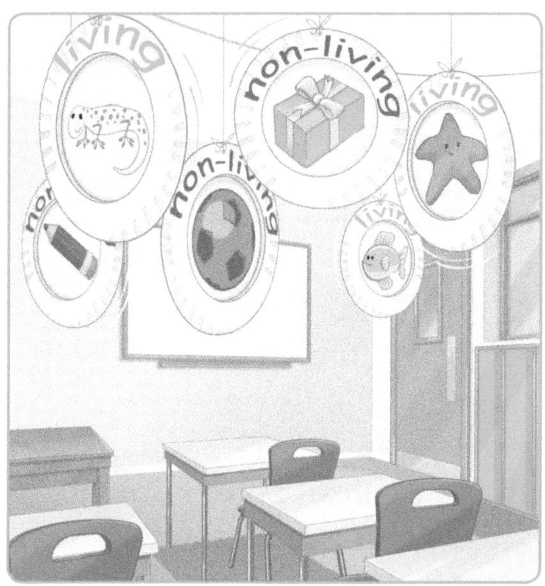

1 a Take a paper plate or circle of card.

b Write 'living' at the top of one side. Write 'non-living' on the other side.

c Cover each side of your plate with pictures of living and non-living things.

- Draw or cut pictures from magazines.
- Make sure you put the pictures on the correct side!
- Label all your drawings.

d Hang your plate so it spins.

2 Look at your classmates' plates. Find two non-living things and two living things that are not on your plate.

Fascinating fact

If you put every different living thing on a separate plate you would need nearly nine million plates!

REAL WORLD SCIENCE

Scientists give a Latin name to every animal and plant species. New species are being discovered so fast that scientists are finding it difficult to find names for them. One day you might be able to help them.

The Latin name of the loggerhead turtle is *Caretta caretta*.

Test yourself

1 List four living things.
2 List four non-living things.
3 Describe two differences between the living and non-living things you have chosen.

5 Moving and growing

Workbook page

Objective

In this section you will describe how living things move and grow.

Animals and plants move in different ways.

Talking point

Talk about the different ways that animals move. Look at the pictures for some clues.

How do plants move?

Mr. Ward does not believe that plants can move. As a scientist you can prove to him that they can move.

1 a Place a flowering plant in a sunny window.
  b Draw a picture to show the direction that the flowers are facing.
2 a At the end of the day, draw a picture to show the direction the flowers are facing now.
3 a Turn the plant so the flowers are facing away from the Sun.
  b After 24 hours, check the plant again.
  c What has happened to the flowers? Draw a picture.
4 a Turn the plant again and leave it for another 24 hours.
  b Check it at the same time the next day. What has happened to the flowers? Draw a picture.

REAL WORLD SCIENCE

Scientists study how animals move to help them to design robots. Some robots can climb walls like spiders do.

Test yourself

1 List four ways that animals can move
2 Describe how plants can move.

6 Feeding and reproducing

Objective

In this section you will describe how living things feed and reproduce.

Different living things need different types of food.

Talking point

Look at the picture of the zoo.

- Imagine you work at the zoo and it is **feeding** time.
- In your group, talk about the foods that each animal will need.

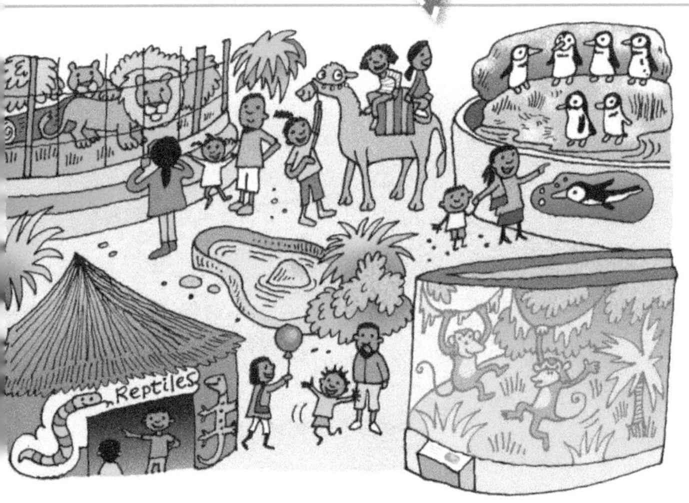

1 a How much of each food do you think you will have to give to the animals?

b Tell the class which foods you have chosen.

Animals and plants need to reproduce to keep the species alive.

You already know that many plants make seeds. These grow to make new plants.

Many animals lay eggs. The young hatch out.

Some animals grow inside their mothers until they are born.

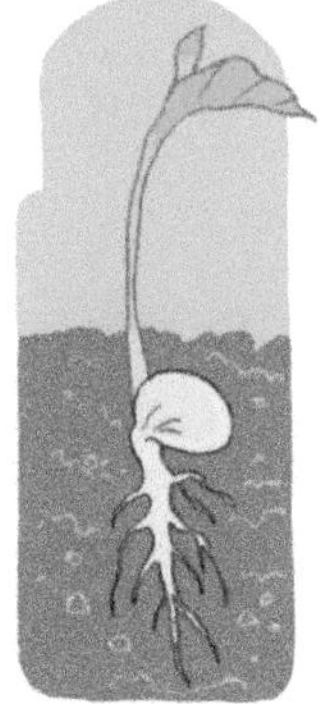

Fascinating fact

The ostrich lays only one egg at a time but hawksbill turtles can lay up to 200 eggs. The record holder might be a fish called the ocean sunfish. This fish lays up to 300 million eggs at one time!

Test yourself

1 Name an animal that:

a hatches from an egg.

b grows inside its mother.

3 Explain why seeds are important.

7 Plants are not all of one type

Workbook page

Objective

In this section you will learn about some different types of plant – trees, shrubs, vines and herbs.

There are many different types of plants. Some important types are:

- trees
- shrubs
- vines
- herbs.

Trees have a single self-supporting stem called a trunk. The trunk is woody and it is usually unbranched for some distance above the ground. Trees are usually taller than other plants.

Shrubs are woody plants. They are smaller than trees and are usually split into separate stems from near the ground. They have many branches. Shrubs are also called bushes.

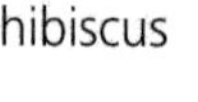

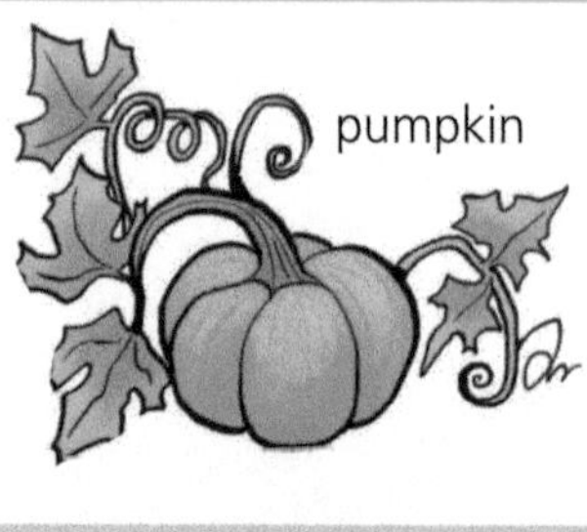

Vines have slender stems that trail or climb. Vines are also called runners.

Herbs are small plants. They have stems that are not woody. Herbs die after flowering and producing seeds. Some herbs are used to flavour foods and to make medicines and perfumes.

Test yourself

1. What is the difference between a tree and a vine?
2. Why don't we use herbs for firewood?
3. What else are shrubs known as?

8 Different plants

Objective

In this section you will identify different kinds of plants.

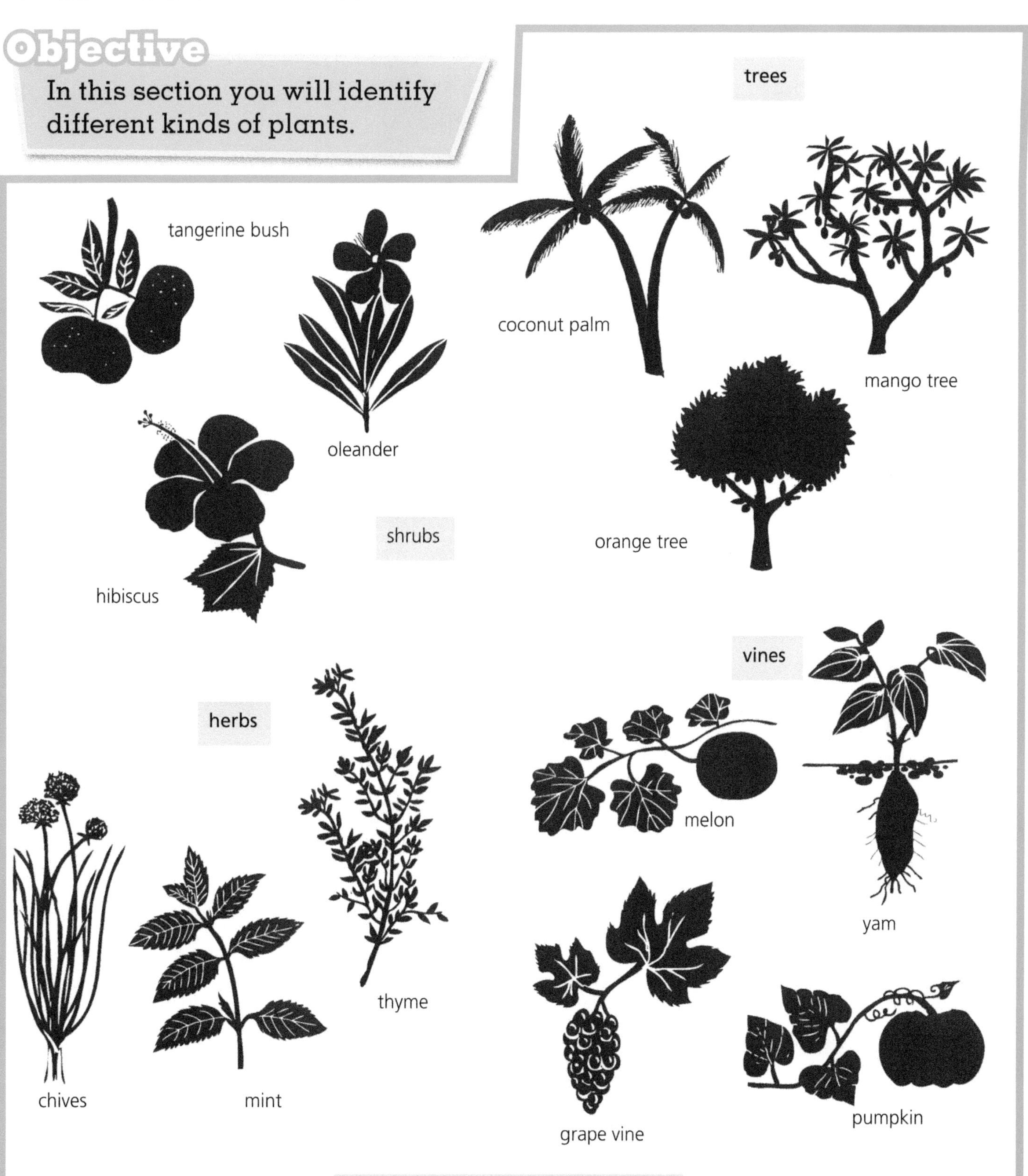

This is one type of identification key.

Plant field trip

Finding different types of plants

Your teacher will take you outside to study some plants.

1 Use the shapes of the plants in the identification key on page 99 to help you.
2 Record the plants you find by drawing them. Note down the location – this is where you find the plants.

Research task

- Find out the names of four trees that grow in your area.
- Draw the shape of these trees so you can help other people to identify them.

REAL WORLD SCIENCE

There are far too many animals and plants on Earth for any scientist to know them all. Scientists use many different keys to help them to identify living things.

1 How many examples of each type of plant did you find?

a trees
b shrubs
c vines
d herbs

Talking point

- Were any of the plants you found difficult to identify?
- Why was the identification difficult?

When we try to group things there are always some that do not clearly belong in one group.

Test yourself

1 What do we use identification keys for?
2 List the four main types of plants.
3 Explain the difference between a tree and a shrub.

9 Leaves are important

Workbook pages 63–64

Objective

In this section you will list some types of leaves and what they are used for.

1 a Which of these leaf shapes have you seen before?

b Can you recognise any of the plants they come from?

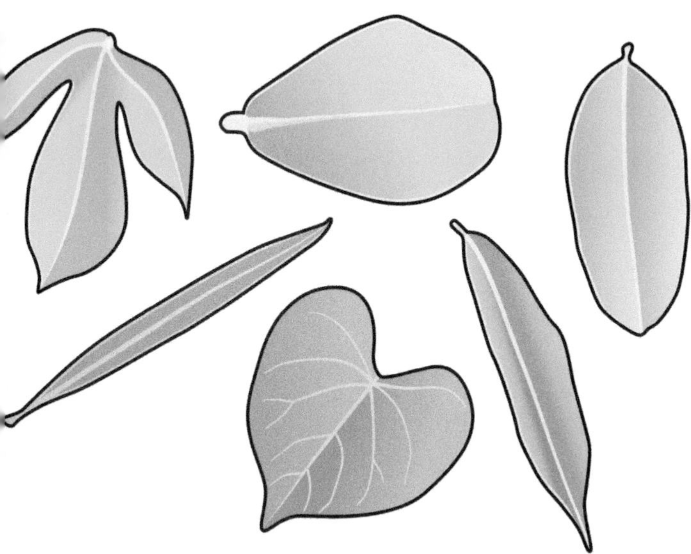

REAL WORLD SCIENCE

Botanists are scientists who study plants. They can check plant leaves to see if a plant is healthy or not. Spotted or lumpy leaves may mean a plant has a disease.

Fascinating fact

Each leaf shape has its own name. Look at these two shapes. Did you find any leaves with these shapes?

palmate pinnate

Collecting and sorting leaves

1 Go outside and find as many different types of leaf as you can.
2 Sort your leaves. Use colour, shape, size, pattern of veins and texture.
3 How many different groups did you make?
4 Share your ideas with the class.

Test yourself

1 List three features that we can use to make groups of leaves.
2 Which was the smallest leaf you collected?
3 Which was the largest leaf you collected?

An exhibition of living and non-living things

Share your living and non-living displays as an exhibition. Your invitations will help to get lots of people to come and see it.

1 Work with your classmates to set up a room so that people can easily see all of your displays.
2 You could make an exhibition guidebook with:
   a information about the displays and what the visitors are going to see
   b a map of the exhibition showing where the visitors can walk.
3 Discuss who is going to do all of the different jobs.
4 Stand by your exhibit ready to answer questions from the visitors.

Tree shadows

1 Choose three of your favourite trees.
   a On black card, draw a picture of each tree. Clearly draw the shape of the tree.
   b Cut out the pictures to make silhouettes of the trees.
   c Close the curtains or blinds to make the room dark.
   d Shine a light onto your silhouettes and make shadows on the wall.
   e Can other people recognise the trees?
2 Different people in the class can make silhouettes of trees, vines, herbs or shrubs.
3 Can other people tell the difference between tree, shrub, vine and herb shadows?

Unit 8: Check your learning so far

Choose the correct picture to answer each question.

1 This person is the tallest.

a 
b 
c 
d 

2 Which two pictures below show living things?

3 Which two pictures below show non-living things?

4 Which picture below shows a thing that was once living?

a 
b 
c 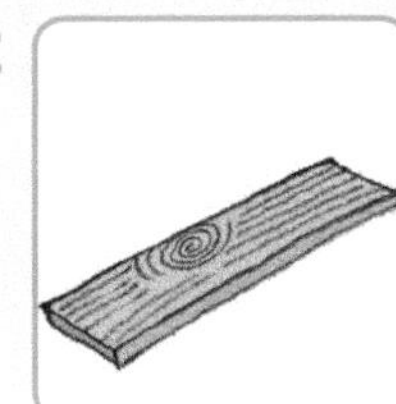
d 
e 

5 Match up the pictures with the labels.

a herb
b tree
c shrub
d vine

6 This is not the shape of a leaf.

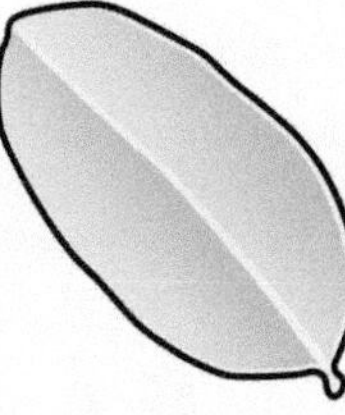
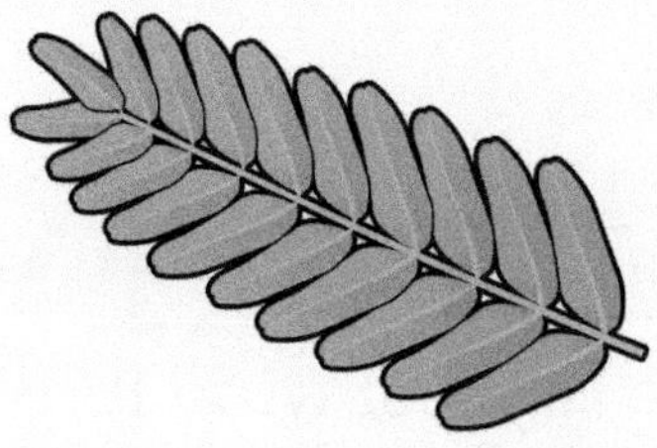

a b c d

Unit 9 • Energy

Electricity, wind and water are all sources of energy.

KEY WORDS
electricity
energy
water
wind

Talking point

- Why is this called a 'wind farm'?
- Why have the wind turbines been placed here?
- What do the turbines do?

Fascinating fact

The biggest wind turbines are almost 100 metres high.

Talking point

- What is the windmill being used for?
- Look at the sails. What does the wind do to the sails?
- How big are the sails?

Fascinating fact

There are water turbine farms under the sea.

Water turbines work like wind turbines. The water turns the turbine instead of the wind.

1 Devices that use energy

Workbook pages

Objective

In this section you will:

- identify devices that use electricity to work
- think about how electric devices help us
- think about why it would be difficult to live without electric devices.

Fascinating fact

People have been using refrigerators in their homes for over 100 years.

Devices help us to carry out tasks more easily. Some devices use **electricity**. Electricity is a form of **energy**.

1 Name an electrical device that you have used today.

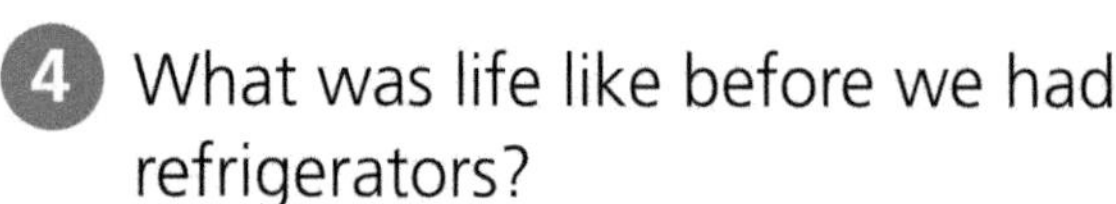

4 What was life like before we had refrigerators?

Talking point

What would your family do if it did not have these devices?

2 Look at the pictures.

a What are these devices called?

b What are they being used for?

c What form of energy do they use?

3 What would life be like without electricity?

Test yourself

1 Write down two devices that you have learnt about.

2 What is each device used for?

2 Old and new devices

Workbook page 67

Objective

In this section you will:
- understand that we use energy to solve problems
- compare old and new technologies
- understand that people keep inventing new things.

Fascinating fact

The first steam trains were introduced to the Caribbean 150 years ago.

This train is powered by steam.

Trains today are powered by diesel and electricity.

Talking point

Why did we stop using steam for trains?

This device uses electricity to get warm. The heat removes creases in our clothes.

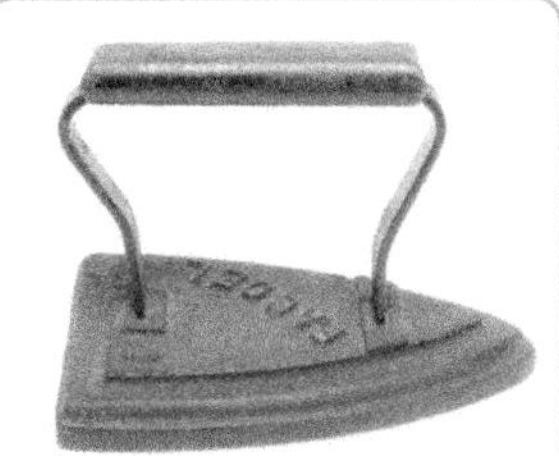

This is an old-fashioned iron.

1. a Was the old-fashioned iron difficult to use?
   b Compare it to the modern iron.

2. Look at the pictures of the boats.
   a Which one is more reliable?
   b Which one do you prefer?

REAL WORLD SCIENCE

Scientists work together to invent new things all the time to make our lives better.

Research task

- Find out about a recent invention.

3 Wind and water are sources of energy

Workbook page

Objective

In this section you will identify and observe devices that use wind and water energy.

Talking point

- Think about your last birthday and how long ago that was.
- Talk about how long ago 4000 years is.

Talking point

- Have you ever seen an old-fashioned water wheel like this one?
- Have you ever felt how strong flowing water can be?
- Have you ever stood under a waterfall?

Fascinating fact

Wind energy has been used for over 4000 years.

Energy from moving air makes the sails on a windmill or wind **turbine** move. This can be used to make electricity or to make machinery work.

A wind turbine.

1 What do you think the machines made?

Water wheels were used to make machines move.

Moving **water** gives energy to paddles to turn machinery or make electricity.

How big do the sails of a windmill need to be?

The sails have to collect as much wind energy as possible.

You will investigate with different sizes of card.

1 **a** Go outside and find a big space.

b Hold the smallest piece of card in front of you and run across the space.

c Repeat this with all the other sizes of card.

2 Compare how difficult it was to run with each size of card.

Running should have been the most difficult with the biggest piece of card.

Your investigation shows that the sails on a windmill need to be big to catch more air. This will capture more energy.

Test yourself

1 What do the sails do on a wind turbine?

2 What is the best size for a wind turbine sail?

a small

b medium

c large

A display of old and new devices

1 Find the oldest device that you can at home.
2 Label the device with its name and what it was used for.
3 Take the old device to school. If your device is too big or fragile, ask if you can take a photograph and bring that to school.
4 Find a picture of a modern device that has replaced the old device that you found.
5 Write a fact file on the old and new devices.
   a Explain how the old device was used and which form of energy it used.
   b Explain how the new device is used and which form of energy it uses.

Researching wind farms

1 Draw a wind turbine.
2 Try to find out the names of the different parts and label them.
3 What do the different parts do?
4 Find out where the nearest wind farm is to you.

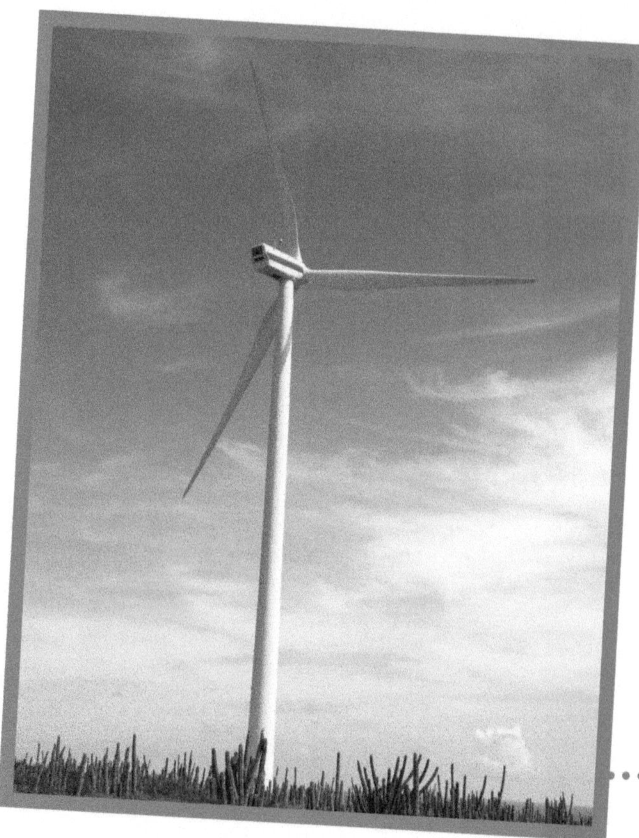

Unit 9: Check your learning so far

Choose the correct picture to answer each question.

1 Which two devices use electricity?

a 

b 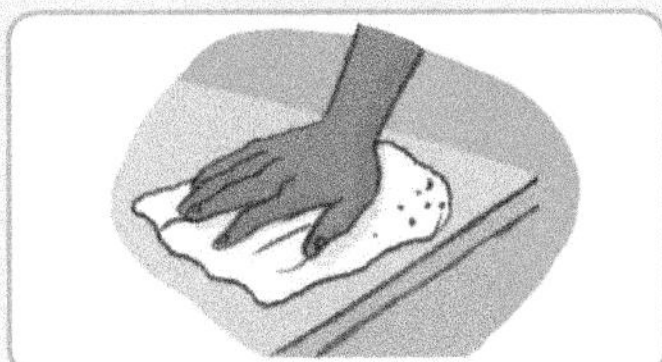

c 

d 

e 

f 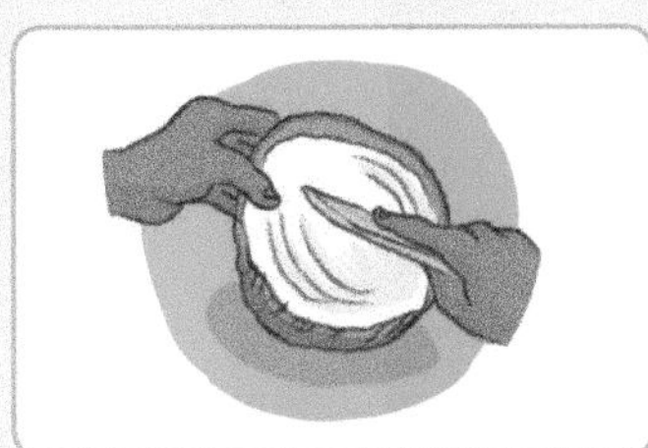

2 Which device is the oldest?

a 

b 

c 

3 Which device uses wind power?

a 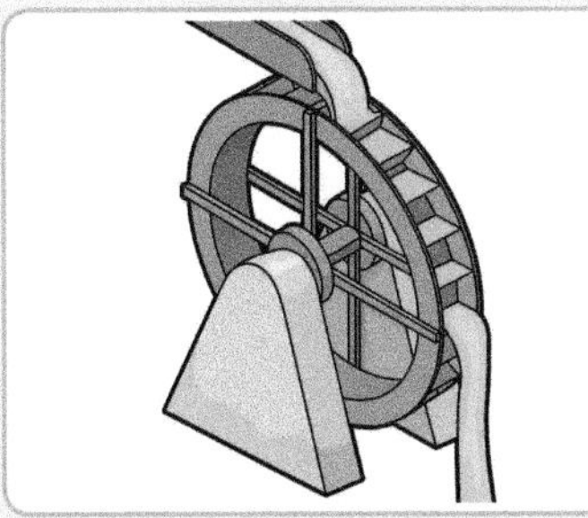

b 

c 

4 Which two devices were used before we had electricity?

a 

b 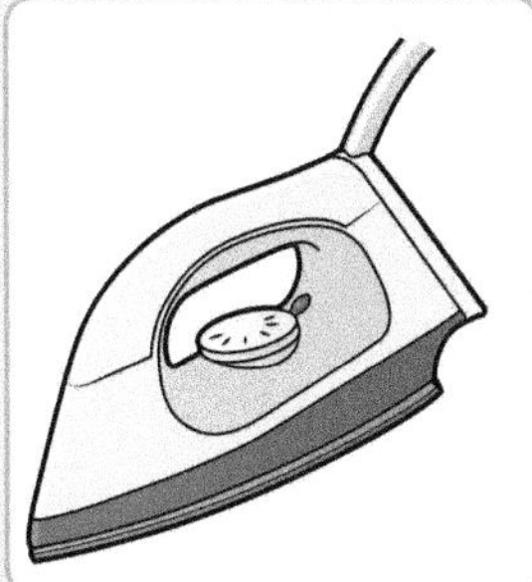

c 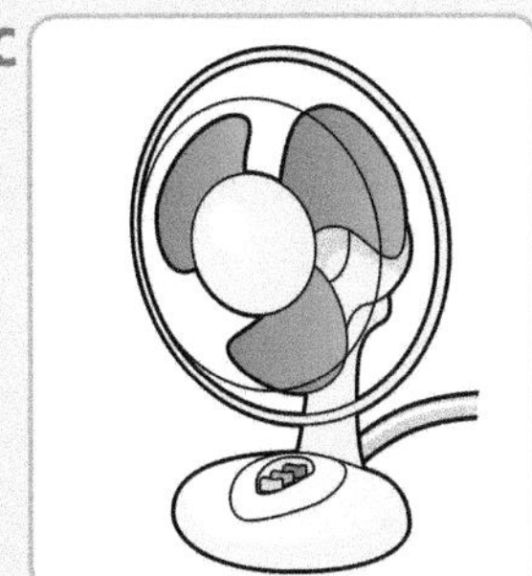

d 

Unit 10 • The Earth's resources

We need to take care of the Earth.

KEY WORDS
air
dust
litter
pollutants

Litter is garbage that has been thrown away carelessly. It is made up of many different items.

Talking point

- Would you drink this water?
- How has the water got so dirty?
- How many different items of litter can you see?
- How can we stop this problem?

Fascinating fact

Between 1989 and 2007, 9.3 million pieces of litter were collected in the Caribbean during the yearly clean-ups.

REAL WORLD SCIENCE

Scientists who work with living things in the sea blame plastic litter for killing marine animals.

Talking point

- Where is the dust coming from?
- How are some people protecting themselves?
- Have you ever been in a dusty place? How did it make you feel?

Fascinating fact

The vacuum cleaner was invented in 1901, more than 100 years ago. Since then there have been many different designs but they all work on the same idea.

Which vacuum cleaner would you prefer to carry around a house?

Dealing with pollution

1 The dangers of litter

Objective

In this section you will review the dangers of litter.

Remember that **litter** is made up of different things.

1 Can you list four different items of litter?

Litter comes from humans. Humans make things to use and then they throw them away. The wind and the sea can carry litter all around the world.

Litter can cause many problems. In the sea old plastic bags and fishing twine can hurt the animals.

2 Look at the picture. Why is the litter a problem here?

Fascinating fact

There is even litter in space. This is old equipment from space travel that has been left in space.

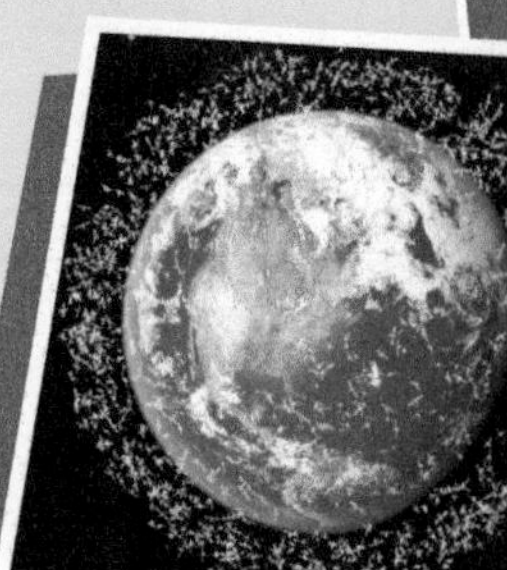

Talking point

Have you ever seen anyone drop litter?

3 a Why do you think some people drop litter?

b Are there enough litter bins?

Finding out how your school manages litter

Carry out a survey of litter bins in your school.

1 How many litter bins did you find?
2 Can you see the bins easily?
3 Were the bins empty?
4 Does the litter stay in the bins or blow out?

2 Litter in schools is a problem

Workbook page 69

Objective

In this section you will discuss how we can prevent the problems of litter.

Fascinating fact

Three out of four people admit that they drop litter.

Talking point

Look at the picture of the bin.

- How does this idea encourage others to put their litter in the bin?
- Have you seen other ideas like this one?

There are many ways to encourage people to use litter bins. Now is your chance to encourage persons to put their litter in the bin.

Encouraging the use of litter bins

Design an interesting and eye-catching structure that other people will want to use.
Think about:

- what equipment you will need
- recycling some of the litter around school
- how big the litter bin will need to be
- how you can stop litter from falling out
- how you can fix the parts of the structure together
- how the litter bin will be emptied
- where you will put it.

Test yourself

1 List four items of litter that you've seen recently.
2 List three dangers of litter.

3 Organising and taking part in a clean-up campaign

Objective

In this section you will start to make your area a cleaner place.

STAY SAFE Check with your teacher before you pick up litter. Never pick up litter with your bare hands.

1 What can you put on your hands to keep them clean while you are picking up litter?

Organising a clean-up campaign

1 Your teacher will take you on a tour of your local area, so you can find out which parts of the area have the most litter.
2 Choose a place that you would like to clean up.
3 Design and make a poster telling people why they should not drop litter.
4 Your teacher will ask local shopkeepers and restaurant owners to display the posters.
5 Go back to the area after a week. Do you see a difference?

Talking point

- Look at the picture of litter in the street.
- How could this litter harm living things?

Research task

- Find out if there are any clean-up programmes in your area.

Fascinating fact

One trillion plastic bags are thrown away every year in the world. Ten million new plastic bags are made every day!

Test yourself

1 Why do we need to stop littering?
2 How does litter affect living things?

4 Air contains pollutants

Workbook page 70

Objective

In this section you will identify at least two air pollutants.

Air can become polluted. The things that make the air dirty and bad are called **pollutants**.

Living things, including humans, need clean air to survive.

Fascinating fact

We breathe in about 1000 litres of air every day.

That is the same amount as 1000 large boxes of juice.

Talking point

- Open your mouth and breathe in. Does the air taste of anything?
- Close your mouth and breathe in through your nose. Do you smell anything?

Clean air has no colour and does not smell or taste of anything. If you can taste or smell anything, the air is polluted.

Not all pollutants in the air are bad for us but many are.

Investigating pollutants in the air

Work with a partner. Your teacher will give you some sealed plastic containers to test.

1 Close your eyes or ask your partner to blindfold you.
2 Open one of the containers and waft the air towards your nose.
   a Can you smell anything?
   b Can you identify the smell?
   c Where might this smell come from?
3 Repeat the test with the other containers.
4 Swap over so you both have a turn.

Test yourself

1 Name two things that polluted the air you tested.

5 Pollutants in the air affect people

Objective

In this section you will discuss how pollutants affect people.

Talking point

What is dust?

Look closely at a beam of light from a bulb or at a beam of sunlight shining through a window.

Can you see the tiny particles glistening? They look like stars, but they are particles of **dust**.

Fascinating fact

Dust contains tiny pieces of our dead skin as well as soil, fibres from carpets and clothing, and many other things.

Dust can make some people feel ill. It can make it difficult for some people to breathe. It can also make some people cough.

Dust affects babies, older people and those with breathing problems such as asthma the most.

Some diseases can travel through the air. When someone with influenza coughs or sneezes, the disease is spread to other people.

This person has influenza, also called the flu. He has a runny nose, a high temperature, a sore throat and a headache.

Talking point

- Talk about how influenza affects
- Have you ever had influenza?
- Did you have the same symptom as the boy in the picture?

Research task

- What other illnesses are caused by air pollution?
- What other diseases can travel through the air?

Test yourself

1 What does clean air taste of?
2 What colour is clean air?

6 Collecting and removing dust from the air

Objective

In this section you will make a trap for dust in the air.

Fascinating fact

Factories have tall chimneys to make the pollution go high above where people are. This is so people don't breathe the polluted air. The tallest factory chimney in the world is 420 metres tall.

Some air purifiers work by pushing air through a filter made of a very fine mesh.

The air passes through the tiny holes in the filter. The tiny dust particles are trapped. This takes the pollution away from the air.

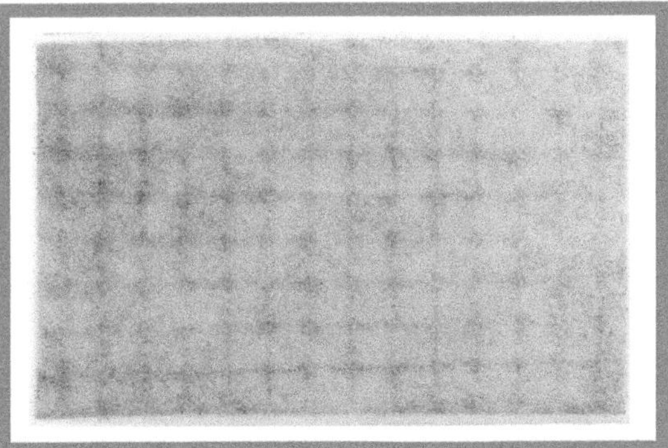

Making an air trap to filter dust from the air

1 a Look at the materials you have been given.
  b Which one would make a good filter?
2 Design an air trap for the window in your classroom.
3 Make your air trap and hang it over an open window.
4 Check the trap at the end of the day. Has the filter collected any dust?

Talking point

Hospitals use air filters. Why do you think they do this?

Test yourself

1 What is dust made from?
2 How can dust affect a person's health?

7 Collecting dust with a dust filter

Workbook pages

Objective

In this section you will compare the amount of air pollution in different areas.

Fascinating fact

Barbecues make lots of air pollution because of the smoke from the charcoal.

Air can be polluted by many things.

Exhaust fumes from vehicles, smoke from burning materials and dust from building work are all air pollutants.

1 What different types of particles can cause air pollution?

Talking point

Which local areas do you think have the most air pollution?

Making a dust filter

1 Work with a partner to make a dust filter.
You will need some cloth to use as a filter, a rubber band and an empty, clean plastic bottle.

Take the top off the plastic bottle.

Squeeze the bottle as hard as you can to squeeze the air out.

Quickly cover the top with a piece of clean white cloth. Put a rubber band around it to hold the cloth securely in place.

2 Release the bottle. This will suck air into the bottle.

3 Look at the filter. Did you collect any dust?
You might need to do this more than once to collect dust.

4 Work with your partner. Compare the amount of dust in different areas.
   a Find a place where you think there will be lots of dust.
   b Use your dust filter to find out how much dust there is.

5 In which place did you collect the most dust? Can you explain why?

8 Protection from air pollution

Objective

In this section you will compare the different devices used to protect workers from air pollution.

Fascinating fact

The air in our homes contains air pollution.

People who build houses work on construction sites where the air is not clean.

They wear masks to stop them breathing in dirty air.

1 Look at the photo of a construction worker's mask.

a Draw what you think the mask would look like if you held it up to the light.

b Make sure you draw the dust particles on the mask.

Firefighters work where there is lots of smoke pollution. Breathing in smoke particles can damage their lungs.

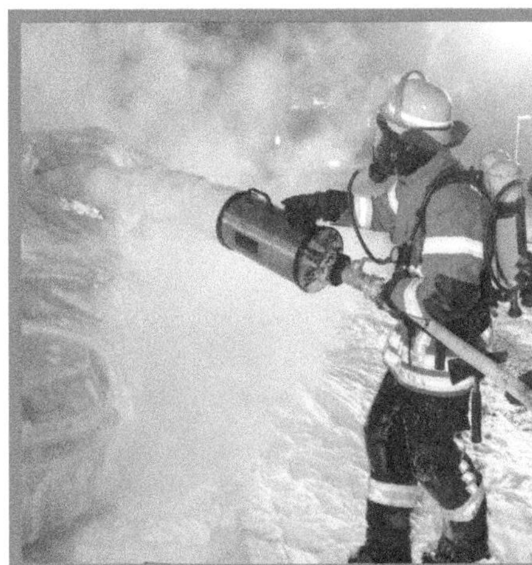

Firefighters have to carry their own air in special tanks.

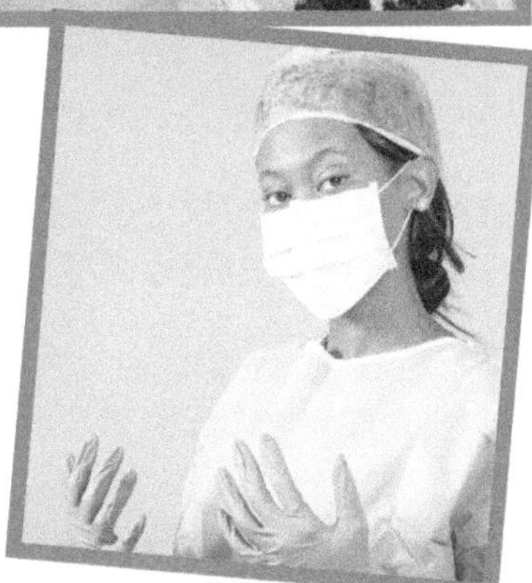

This doctor works with people who are ill. She is wearing a mask to stop diseases getting into her lungs.

2 a Write down the name of an illness that can travel from person to person through the air.

b Write down the name of an illness that is caused by polluted air.

Research task

- Find out the name of an illness that can be made worse by polluted air.

Test yourself

1 Why is it important to have clean air?

Air pollution poster

Make a poster to tell other people about air pollution.

1 Tell them where air pollution comes from.
2 Give two examples of air pollution.
3 Inform them how they can stop the pollution making them ill.

Making an air filter mask

1 Think of a job where the worker needs to wear a mask.
2 Design and make a mask for this worker to wear.
   a Which material will you use?
   b How big does the mask need to be?
   c How will you attach the mask to the worker's face?
   d Does the mask have to cover the person's nose, mouth or both?

Unit 10: Check your learning so far

Choose the correct picture for each clue.

1. Litter is made from this.
2. This is an air pollutant.
3. This device removes dust.
4. This device is used to measure how much dust there is.
5. This device protects workers from air pollution.

a

b

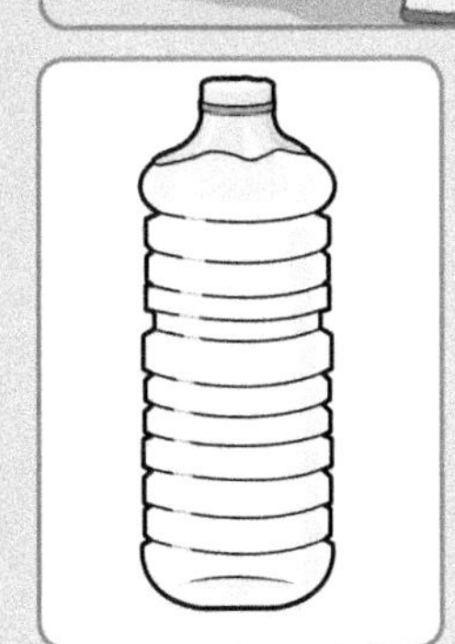

c

d

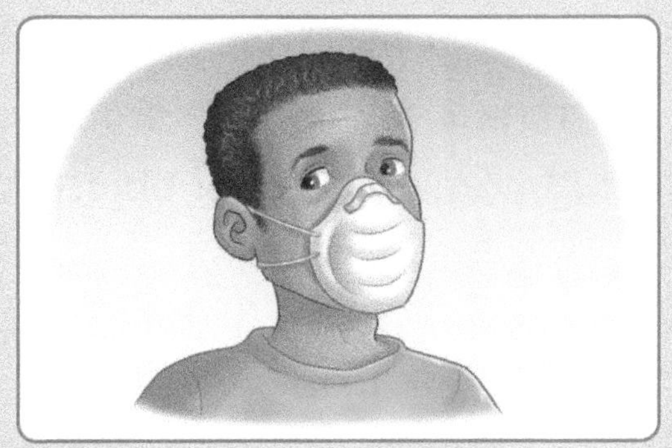

e

Glossary

A

adapted When the bodies of animals and plants are matched to the place that they live in.

adult A fully grown animal or plant.

air An invisible gas that is all around us.

alike When two or more plants or animals look the same.

anemometer A device that is used to measure the speed of wind.

axis An imaginary line that runs through the centre of an object that spins. The Earth rotates on its axis.

C

carnivore An animal that eats only meat.

clay A type of soil that is sticky when it is wet. It can be made into different shapes.

condensation When water vapour changes from a gas to a liquid. This happens when damp air meets something cold.

consumer An organism that feeds on other organisms in a food chain.

D

develop When an animal or plant grows and changes.

device A tool that does a special job.

different When two or more things are not the same.

dust Small pieces of objects that float in the air.

E

Earth The planet we live on. This can also be another word for soil.

egg The object covered in a shell that some female animals lay, which is the beginning of a new animal.

electricity A flow of energy, which often comes from a battery or the mains.

energy When something is able to do work, for example carrying an object.

environment The surroundings that an animal or plant lives in, which affects how it lives.

evaporation When a liquid turns into a gas.

F

feeding When animals eat other plants or animals.

G

grow When a plant or animal gets bigger.

growth The process of a living thing getting bigger.

H

habitat The place where an animal or plant lives.

hardness When something cannot be easily broken, cut or squashed.

herbivore An animal that only eats plants.

L

larva The caterpillar stage of some insects.

liquid Something that can flow and be poured.

litter Human-made things that are left on the ground, for example paper and cans.

living When an animal or plant is alive.

M

mechanical Energy that is made by or from a machine.

Moon The round object that can be seen in the sky at night and that moves around the Earth.

motion When an object moves or is moved.

N

non-living When something is not alive. It is not a living thing.

O

omnivore An animal that eats both plants and animals.

organ A group of tissues that form together to make a specific structure in the body. It carries out special functions.

P

phases of the Moon   Different shapes of the Moon that can be seen.

physical characteristic   How something looks, for example eye colour or the shape of a nose.

planet   An object made of rock or gas in the Solar System.

pollutant   Something that makes the environment dirty.

pollution   An activity that puts harmful materials into the environment.

predator   An animal that hunts and eats other animals.

prey   An animal that is hunted and eaten by another animal.

producer   Green plants that use the light of the Sun to make food in their leaves.

pull   A power that moves objects towards you.

pupa   When an insect grows and it is something between a larva and an adult.

push   A power that moves objects away from you.

R

rain gauge   Something that measures how much rain falls.

reproduce   When living things make new ones of themselves, for example humans make babies.

rubber   A material that is stretchy.

S

seed   The small, hard part of a plant that grows into a new plant.

solar energy   The energy that comes from the Sun.

Solar System   The Sun, and the eight planets and other objects in space that orbit the Sun.

solid   A material that keeps its shape and is hard to squash.

solid waste management   Removing and treating solid waste made by a community such as a village or town.

start   To begin something.

stop   To end something.

strong   When something cannot be broken easily.

Sun The star that the Earth moves around and that gives us all our warmth and light.

survive To carry on living.

T

turbine A machine that converts one form of energy into another to make electricity. It is usually has vanes on it to make it move fast in wind or water.

V

vibrate To constantly move backwards and forwards.

W

waste Materials and objects that are not wanted anymore and are thrown away.

water A liquid that has no taste or colour. All animals and plants need water to live.

water cycle The never-ending way that water turns into clouds and rain.

wind The natural movement of air.